Digital and
Kalman filtering

Digital and Kalman filtering

An introduction to discrete-time filtering and optimum linear estimation

Second Edition

S. M. Bozic

School of Electronic and Electrical Engineering
University of Birmingham

Halsted Press
An imprint of John Wiley & Sons, Inc.
New York   Toronto

© 1994 S. M. Bozic

First published in Great Britain 1994

Library of Congress Cataloging-in-Publication Data

Available upon request

ISBN 0 470 23401 6

Printed and bound in Great Britain.

Preface to the first edition

The availability of digital computers has stimulated the use of digital signal processing (or time series analysis) in many diverse fields covering engineering and also, for example, medicine and economics. Therefore, the terms digital filtering, Kalman filtering, and other types of processing, appear quite often in the present-day professional literature. The aim of this two-part book is to give a relatively simple introduction to digital and Kalman filtering. The first part covers a filtering operation as normally understood in electrical engineering and specified in the frequency-domain. The second part deals with the filtering of noisy data in order to extract the signal from noise in an optimum (minimum mean-square error) sense.

Digital filtering, as used in the title, refers to part 1 of the book, but the subtitle specifies it more closely as an introduction to discrete-time filtering which is a common theory for digital or other types of filter realizations. The actual realization of discrete-time filters, not discussed here, can be done either in sampled-data form (using bucket-brigade or charge coupled devices), or in digital form (using binary logic circuitry). Alternatively, discrete-time filters, described in terms of difference equation algorithms, can be handled on digital computers.

Kalman filtering, as used in the title, refers to part 2 of the book, and again the subtitle describes it more closely as an introduction to linear estimation theory developed in discrete-time domain. Although this part deals initially with some digital filter structures, it develops its own terminology. It introduces the criterion of the minimum mean-square error, scalar and vector Wiener and Kalman filtering. However, the main and most practical topic is the Kalman filtering algorithm which in most applications requires the use of digital computers.

Most of the material has been used by the author in postgraduate courses over the past five years. The presentation is in 'tutorial' form, but readers are assumed to be familiar with basic circuit theory, statistical averages, and elementary matrices. Various central topics are developed gradually with a number of examples and problems with solutions. Therefore, the book is suitable both for introductory postgraduate and undergraduate courses.

The author wishes to acknowledge helpful discussions with Dr J. A. Edwards of this School, who also helped with some of the computer programs.

SMB
1979

v

Preface to the second edition

Over a period of time a number of communications have been received about the first edition of this book. Some were comments about contents and others were corrections regarding the examples and problems. These have been taken into account in the second edition, and the author wishes to thank all the readers for their contributions.

In the second edition some material has been reorganized and some new topics have been added, In *chapter two*, the main new addition is a section on the graphic method for frequency response computations. In *chapter three*, some sections have been rearranged, and also two new sections added. One deals with the FIR filter design by the frequency sampling method and another one introduces the equiripple FIR filter design. In *chapter four*, more information is given about the frequency transformations, quantization effects, and also the wave digital filters have been introduced. In *chapter five*, two new sections have been added: circular convolution, and an introduction to multirate digital filters.

In part 2, *chapters six and seven* have been considerably rearranged. New material has been added in chapter six: IIR Wiener filters, and adaptive FIR filters. Also, in both chapters Wiener and Kalman filters have been interpreted as lowpass filters with automatically controlled cut-off, hence behaving as intelligent filters. In *chapter eight*, a new section has been added dealing with practical aspects of Kalman filtering. Similarly, *chapter nine* has been changed by replacing the vector Wiener filter example with two new examples showing practical aspects in setting up the stage for Kalman filter application.

The author wishes to thank referees for their valuable suggestions. It would be nice, in future, to link the examples and problems to the Matlab software package.

SMB
1994

Contents

Contents

Part 1 – Digital filtering

Introduction

Digital filtering is used here as a well-established title, but with the reservation that we are dealing only with time sequences of sampled-data signals. However, the fundamental theory presented for discrete-time signals is general and can also be applied to digital filtering.

It is important to clarify the terminology used here and in the general field of signal processing. The analogue or continuous-time signal means a signal continuous in both time and amplitude. However, the term continuous-time implies only that the independent variable takes on a continuous range of values, but the amplitude is not necessarily restricted to a finite set of values, as discussed by Rabiner (1). Discrete-time implies that signals are defined only for discrete values of time, i.e. time is quantized. Such discrete-time signals are often referred to as sampled-data or analogue sample signals. The widely-used term digital implies that both time and amplitude are quantized. A digital system is therefore one in which a signal is represented as a sequence of numbers which take on only a finite set of values.

It is also important to clarify the notation used here. In mathematics the time increments or decrements are denoted by Δt, but in digital filtering the sampling time interval T is generally used. A sample of input signal at time $t = kT$ is denoted as $x(k)$, where T is neglected (or taken as unity) and k is an integer number; similarly, for the output we have $y(k)$. In many papers and textbooks, particularly mathematical ones (difference equations), the notation is x_k, y_k. We use here the notation $x(k)$, $y(k)$ for the following reasons:

(i)   it is a direct extension of the familiar $x(t)$, $y(t)$ notation used for the continuous-time functions;
(ii) it is suitable for extension to the state-variable notation, where the subscripts in $x_1(k)$, $x_2(k)$ . . . refer to states;
(iii) it is more convenient for handling complicated indices, for example $x(N-\frac{1}{2})$.

It will be seen later that digital filtering consists of taking (usually) equidistant discrete-time samples of a continuous-time function, or values of some discrete-time process, and performing operations such as discrete-time delay, multiplication by a constant and addition to obtain the desired result.

The first chapter introduces discrete-time concepts using some simple and familar analogue filters, and also shows how a discrete-time description can arise directly from the type of operation of a system, e.g. radar tracking. The z-transform is then introduced

1

as a compact representation of discrete-time sequences, and also as a link with the Laplace transformation. The second chapter expands the basic concepts established in the first chapter, dealing first with the time response of a discrete-time system (difference equations). Then we introduce and discuss the transfer function, inversion from z-variable back to time-variable, and the frequency response of a digital filter. This chapter ends with the realization schemes and classification of digital filters into the basic nonrecursive and recursive types, whose design techniques are presented in the third and fourth chapters respectively. In both cases design examples and computer calculated responses are given to illustrate various design methods. In the fifth chapter we return to some of the basic relationships introduced in the first chapter, and deal with two important topics in discrete-time processing. First we develop the discrete Fourier series representation of periodic sequences, which enables formulation of the discrete Fourier transform (DFT) for aperiodic finite sequences. The second topic is the inverse filter, which is an important concept used in many fields for removal or reduction of undesirable parts of a sequence. The concept of minimum error energy is introduced as an optimization technique for the finite length inverse filter coefficients. At the end of each chapter a number of problems is given with solutions at the end of the book.

It is of interest to mention that the approach used in this book is a form of transition from the continuous- (or analogue) to discrete-time (or digital) systems, since most students have been taught electrical engineering in terms of continuous-time concepts. An alternative approach is to study electrical engineering directly in terms of discrete-time concepts without a reference to continuous-time systems. It appears that the discrete case is a natural one to the uninhibited mind, but special and somewhat mysterious after a thorough grounding in continuous concepts, as discussed by Steiglitz (2), p. viii.

1

Introduction to discrete-time filtering

1.0 Introduction

In continuous-time, the filtering operation is associated with RC or LC type of circuits. Therefore, in the first section of this chapter, we consider two simple filtering circuits (RC and RLC) described in continuous-time by differential equations, and we find their discrete-time equivalents, i.e. difference equations. There are also situations in which difference equations are obtained directly, as illustrated in section 1.2. In continuous-time we usually represent differential equations in the complex frequency s-domain by means of Laplace transformation, and from these we obtain the frequency response along the $s = j\omega$ axis. Similarly, difference equations in discrete-time are transformed into z-domain using z-transformation which is briefly introduced in the third section of this chapter. The relationship between z and s is then established in section 1.4.

1.1 Continuous and discrete-time analysis

We are all familiar with the description of continuous-time dynamic systems in terms of differential equations. As an introduction to the discrete-time description and the process of filtering we consider a few differential equations and transform them into their discrete-time equivalents, i.e. difference equations.

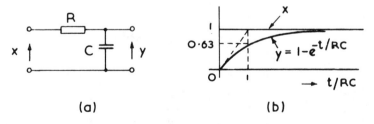

(a)   (b)

Fig. 1.1   (a) Simple first-order RC filter; (b) solution for unit step input

Consider the typical first-order RC filter in fig. 1.1(a) where x and y represent the input and output voltages respectively. For this simple network x and y are related by the differential equation

$$RC\frac{dy}{dt} + y = x \tag{1.1}$$

3

The solution, for a unit step input and zero initial condition, is shown in fig. 1.1(b). To derive the discrete-time equivalent for equation 1.1, we use the method of backward differences, described by Hovanessian *et al.* (3), and obtain

$$RC\frac{y_k - y_{k-1}}{\Delta t} + y_k = x_k$$

where we have used the standard mathematical notation. Solving for y_k, we have

$$y_k = \frac{1}{1 + \Delta t/RC} y_{k-1} + \frac{\Delta t/RC}{1 + \Delta t/RC} x_k$$

Using the approximation $(1 + \Delta t/RC)^{-1} \simeq 1 - \Delta t/RC$, where we have neglected terms of higher order, we obtain

$$y(k) = a_0 x(k) + b_1 y(k-1) \tag{1.2}$$

with $a_0 = \Delta t/RC$ and $b_1 = 1 - a_0$. Note that we have changed to sample notation, as discussed in the introduction, with k representing the discrete integer time parameter instead of $t = k\Delta t$. The above result enables us to draw fig. 1.2(a) which is the

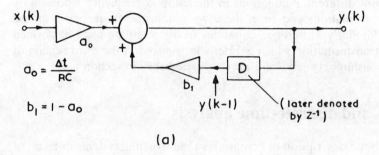

(a)

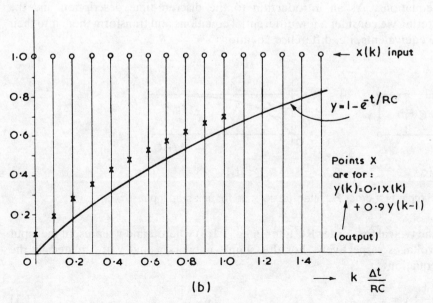

(b)

Fig. 1.2   (a) Discrete-time representation of fig. 1.1(a); (b) solution for unit step input

discrete-time equivalent of the continuous-time system shown in fig. 1.1(a). In fig. 1.2(a), triangles are used to represent multiplication by the factor written beside them, rectangles to represent delay units denoted also by $D(=\Delta t)$, and circles to represent addition. Choosing numerical values $a_0 = \Delta t/RC = 0.1$, $b_1 = 1 - a_0 = 0.9$, and $y(-1) = 0$ as the initial condition, we obtain fig. 1.2(b), where the first ten points have been calculated using a slide rule.

A typical second-order differential equation is

$$\frac{d^2y}{dt^2} + 2\sigma\frac{dy}{dt} + \omega_0^2 y = \omega_0^2 x \tag{1.3}$$

where again x and y refer to the input and output respectively. Such an equation describes, for example, the LRC circuit in fig. 1.3(a) with $\sigma = R/2L$ and $\omega_0^2 = 1/LC$. The

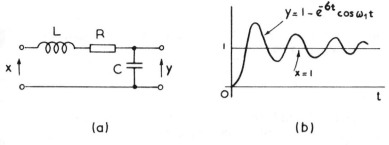

Fig. 1.3 (a) LRC filter; (b) solution for unit step input

solution for the unit step input is given in fig. 1.3(b) for the underdamped case when $\omega_1 = \sqrt{(\omega_0^2 - \sigma^2)} \gg 1$. The discrete-time form of equation 1.3 can be shown to be

$$y(k) = a_0 x(k) + b_1 y(k-1) + b_2 y(k-2) \tag{1.4}$$

where backward differences have been used, and the coefficients are functions of σ and ω_0 (see section 1 of the appendix). The discrete-time operational scheme for the second-order difference equation 1.4 is shown in fig. 1.4, with the same notation as in fig.

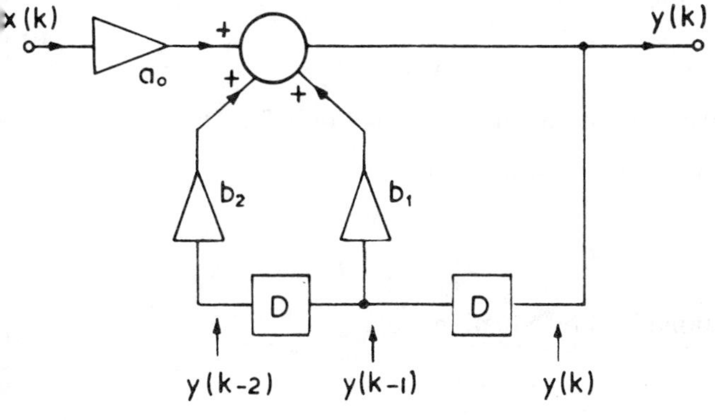

$$a_0 = 1, \quad b_1 = 2(1 + \sigma\Delta t)/\omega_0^2 \Delta t^2, \quad b_2 = -1/\omega_0^2 \Delta t^2$$

Fig. 1.4 Discrete-time representation of fig. 1.3(a)

1.2(a). The unit step response in discrete-time has not been calculated, but the interested reader may do it as an exercise.

The input $x(t)$ and output $y(t)$ variables in equations 1.1 and 1.3 are related in terms of differential equations. Another way of expressing their relationship, in continuous-time systems, is in terms of the convolution integral

$$y(t) = \int_0^t h(\tau) x(t - \tau) \, d\tau \tag{1.5}$$

where $h(\tau)$ represents the impulse response, and $x(t)$ is an arbitrary input signal. The discrete-time form of the above equation is

$$y(k) = \sum_{i=0}^{k} h(i) x(k - i) \tag{1.6}$$

known as the convolution summation, often written as $y = h * x$ (note that $h * x = x * h$). We can illustrate in a simple way that equation 1.6 is correct by considering equation 1.2 first for the impulse (i.e. unit-sample) input,

$$x(k) = \delta(k) = \begin{cases} 1 & k = 0 \\ 0 & k \neq 0 \end{cases}$$

and then for an arbitrary input $x(k)$. In both cases we assume zero initial conditions.

For the impulse input, equation 1.2 produces the sequence

$$
\begin{aligned}
y(0) &= h(0) = a_0 \\
y(1) &= h(1) = a_0 b_1 \\
y(2) &= h(2) = a_0 b_1^2 \\
&\vdots \\
y(k) &= h(k) = a_0 b_1^k
\end{aligned} \tag{1.7}
$$

but for an arbitrary input signal $x(k)$, we obtain

$$
\begin{aligned}
y(0) &= a_0 x(0) \\
y(1) &= a_0 x(1) + a_0 b_1 x(0) \\
y(2) &= a_0 x(2) + a_0 b_1 x(1) + a_0 b_1^2 x(0) \\
&\vdots \\
y(k) &= a_0 x(k) + a_0 b_1 x(k-1) + \ldots + a_0 b_1^k x(0)
\end{aligned}
$$

Comparing the above equation with the set of values in equation 1.7 we have

$$y(k) = h(0) x(k) + h(1) x(k-1) + \ldots + h(k) x(0)$$

or $$y(k) = \sum_{i=0}^{k} h(i) x(k - i)$$

which is the convolution summation given by equation 1.6.

1.2   Discrete-time processing in radar tracking

This is an example of direct formulation of the problem in discrete-time, and a simplified diagram, fig. 1.5(a), shows the main points to be discussed. A radar beam is used to

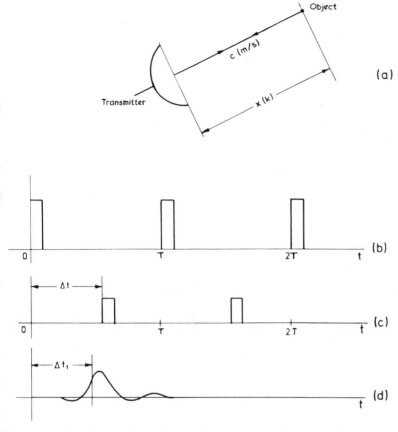

Fig. 1.5   (a) Simplified radar tracking system; (b) ideal transmitted pulses; (c) ideal received pulses; (d) typical received pulses

determine the range and velocity of an object at a distance x from the transmitter. Fig. 1.5 shows the set of ideal transmitted and received pulses together with a typical received pulse. The information required is the value of the time interval Δt representing the time passed for the radio wave to travel to the object and back. The typical received signal is not of ideal shape due to various disturbances, and we measure $\Delta t_1 \neq \Delta t$. The range estimate $x = c \Delta t_1/2$ from one measurement can therefore cause large errors; c is the pulse propagation in space. To reduce the error, a periodic sequence of pulses is transmitted every T seconds, as indicated in fig. 1.5, which produces a sequence of measured values of range $x(0), x(1), \ldots, x(k)$. In many cases the object is moving and its velocity (rate of change of range) is required, together with the object's range at a time one radar pulse in the future.

To establish a processing scheme for radar data we introduce the following quantities:

$x(k)$, the measurement of the object's range obtained from the kth radar pulse return;
$y(k)$, the estimate of the object's range at the kth radar pulse after data processing;
$\dot{y}(k)$, the estimate of the object's velocity at the kth radar pulse after data processing;
$y_p(k)$, the prediction of the object's range at the kth radar pulse, obtained at the $(k-1)$th radar pulse after data processing.

The last of the above quantities can be expressed as

$$y_p(k) = y(k-1) + T\dot{y}(k-1)$$

where T is the time interval between transmitted pulses. The next relationship is established in the following way:

$$y(k) = y_p(k) + \alpha[x(k) - y_p(k)]$$

where the predicted range is corrected by the error between the measured and predicted values, which is scaled by factor $\alpha > 0$. In a similar way, we have for the velocity

$$\dot{y}(k) = \dot{y}(k-1) + \frac{\beta}{T}[x(k) - y_p(k)] \qquad \text{where } \beta > 0$$

The set of relationships formed from the above equations

$$\left. \begin{aligned} y_p(k) &= y(k) - 1) + T\dot{y}(k-1) \\ y(k)_p &= y_p(k + \alpha[x(k) - y_p(k)] \\ \dot{y}(k) &= \dot{y}(k-1) + \frac{\beta}{T}[x(k) - y_p(k)] \end{aligned} \right\} \tag{1.8}$$

describes a signal processing scheme known as the alpha–beta $(\alpha$–$\beta)$ tracking equations. The hardware structure defined by these equations is shown in fig. 1.6. The input $x(k)$

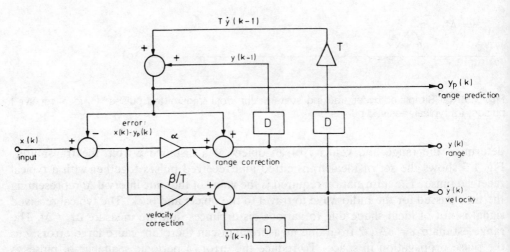

Fig. 1.6 α–β tracker processor

represents the measured data, and the three outputs of the processing unit, $y(k)$, $\dot{y}(k)$ and $y_p(k)$ represent the range, velocity and range prediction respectively.

More detail on the α–β tracking can be found in Cadzow (4), but the above set of equations is further considered in problems 1.6 and 1.7, and later in problem 2.12.

The block diagrams shown in figs 1.2(a), 1.4 and 1.6 are digital filters or processors represented in the hardware form. Their computational algorithms, given by equations 1.2, 1.4 and 1.8 respectively, are often used as such in time series analysis, for example in Anderson(5). However, in engineering practice it is usually preferable to work in the frequency-domain, not only because such representation shows clearly the type of filtering, but because design and measurements are often more precise in the frequency-domain than in the time-domain.

We know that in continuous-time systems the time-domain is transformed into the complex frequency (s-) domain using the Laplace transform. Similarly, in discrete-time systems we can use the z-transform which enables analysis of these systems in the frequency-domain. An introduction to z-transform and its properties is given in the following two sections.

1.3   z-transform

Let $f(k)$ represent any discrete-time variable, i.e. $x(k)$ or $y(k)$ in the examples of sections 1.1 and 1.2, or any other discrete-time variable obtained for example from data recording. This discrete-time variable is a time sequence which can be written, for example, as

$$[f(k)] = f(0), f(1), f(2), \ldots, f(k), \ldots$$

taking place at times

$$0, \Delta t, 2\Delta t, \ldots, k\Delta t, \ldots$$

A useful way of representing this data is in the form of a polynomial

$$f(0) + f(1)z^{-1} + f(2)z^{-2} + \ldots + f(k)z^{-k}$$

which is a power series in z^{-k} having as coefficients the values of the time sequence $[f(k)]$. Alternatively, this can be written in the following compact form (see, for example, Cadzow(4) and Freeman(6)):

$$Z[f(k)] = F(z) = \sum_{k=0}^{\infty} f(k)z^{-k} \tag{1.9}$$

where we have assumed $f(k) = 0$ for $k < 0$. Such a function is known as the z-transform of the $f(k)$ sequence. The z-transform gives a procedure by which a sequence of numbers can be transformed into a function of the complex variable z. The variable z^{-k} can be interpreted as a type of operator that upon multiplication shifts signal samples to the right (delays) by k time units. Multiplication by z^k shifts signals to the left. Another interpretation of the variable z is developed in the following section, where it is shown that z, in sampled-data systems, is a complex number such that $z = e^{sT}$, where T ($= \Delta t$ in the notation used up to here) is the sampling time interval.

There are various methods of finding closed form expressions for the z-transforms of common time sequences, which are discussed by Cadzow(4) and Freeman(6). As an illustration, we consider the case of the geometric sequence (or exponential signal) given by $f(k) = c^k$ for $k \geq 0$, and $f(k) = 0$ for $k < 0$. The z-transform of this sequence is given by

$$Z[f(k)] = F(z) = \sum_{k=0}^{\infty} c^k z^{-k}$$

This is a geometric series whose sum is given by

$$S = A\frac{1 - r^k}{1 - r}$$

where $r = cz^{-1}$, $A = 1$, and $k \to \infty$, and we can write

$$Z[c^k] = \frac{1}{1 - cz^{-1}} \qquad (1.10)$$

for $|r| = |cz^{-1}| < 1$ or $|z| > |c|$. The following three cases are typical sequences whose z-transforms can be obtained from equation 1.10. First, for $c = 1$, we have

$$Z[1] = \frac{1}{1 - z^{-1}} \qquad (1.11)$$

for $|z| > 1$, which is the z-transform of the constant sequence (or sampled unit step). Similarly, by letting $c = ae^{jb}$, we have

$$Z[a^k e^{jbk}] = \frac{1}{1 - ae^{jb} z^{-1}} \qquad (1.12)$$

for $|z| > a$. From equation 1.12 we can easily obtain $Z[a^k \cos bk\,\Delta t]$ and $Z[a^k \sin bk\,\Delta t]$, as in Table 1.1.

Table 1.1   Commonly used z-transforms

$x(k), k \ge 0$		$X(z)$
Unit pulse	$\delta(k)$	1
Unit step	1	$1/(1 - z^{-1})$
Exponential	c^k	$1/(1 - cz^{-1})$
Ramp	k	$z^{-1}/(1 - z^{-1})^2$
Cosine wave	$a^k \cos bkT$	$(1 - az^{-1} \cos bT)/(1 - 2az^{-1} \cos bT + a^2 z^{-2})$
Sine wave	$a^k \sin bkT$	$az^{-1} \sin bT/(1 - 2az^{-1} \cos bT + a^2 z^{-2})$

Furthermore, we can obtain the z-transform for the sampled ramp signal $x(k) = k$, for $k \ge 0$, from equation 1.10 as follows. Writing equation 1.10 as

$$\sum_{k=0}^{\infty} c^k z^{-k} = \frac{z}{z - c}$$

and differentiating both sides with respect to z,

$$z^{-1} \sum_{k=0}^{\infty} kc^k z^{-k} = \frac{c}{(z - c)^2}$$

Multiplying by z and setting $c = 1$, we have

$$\sum_{k=0}^{\infty} kz^{-k} = \frac{z}{(z - 1)^2} \qquad (1.13)$$

which is also given in Table 1.1.

We consider next the case of a signal delayed by i discrete-time units. Its z-transform is given by

$$Z[f(k - i)] = \sum_{k=0}^{\infty} f(k - i) z^{-k}$$

or $Z[f(k-i)] = z^{-i} \sum_{k} f(k-i) z^{-(k-i)}$

$$= z^{-i} \sum_{m=-i}^{\infty} f(m) z^{-m}$$

which results in

$$Z[f(k-i)] = z^{-i} F(z) \tag{1.14}$$

where we have assumed zero initial conditions $f(m) = 0$ for $m < 0$ (for nonzero conditions, see, for example, Cadzow(4), pp. 162–4). This is an important result, also known as the right-shifting property, which will be used later in various sections. The above relationship corresponds to the Laplace time-shift theorem

$$L[f(t-\tau)] = e^{-\tau s} F(s) \tag{1.15}$$

Derivations of many other z-transform relationships can be found in Cadzow(4), Freeman(6) and Jury(7). In Table 1.1 we have used T in the last two entries instead of Δt. This is the notation which is generally accepted and which will be used from the next section onwards.

We shall be using the z-transform as defined by equation 1.9. However, mathematicians and some control engineers use the Laplace definition of z-transform (described by Robinson(8), p. 341) given by

$$F(z) = \sum_{k=0}^{\infty} f(k) z^k$$

1.4   *z*-transform relation to the Laplace transform

In communications and control practice, the discrete-time sequences $x(k)$ and $y(k)$, or in the general notation $f(k)$, are samples of a continuous-time waveform. These can be interpreted as a form of modulation of the signal $f(t)$ by a sequence of impulses as in fig. 1.7. A signal $f(t)$ is sampled every T seconds, and the sampled output wave

$$f_s(t) = f(t) x f_1(t)$$

can be represented as

$$f_s(t) = f(t) \sum_{k=0}^{\infty} \delta(t-kT)$$

or $f_s(t) = \sum_{k=0}^{\infty} f(kT) \delta(t-kT) \tag{1.16}$

where the subscript s denotes the sampled signal. (In control systems analysis, sampled waveforms are commonly denoted by an asterisk.) The notation in this section is changed from the one used previously in two ways: $f(k)$ is changed to $f(kT)$, and Δt is changed to T. This notation is widely used in the field of digital filters.

The Laplace transform of $\delta(t-kT)$ is

$$L[\delta(t-kT)] = e^{-skT}$$

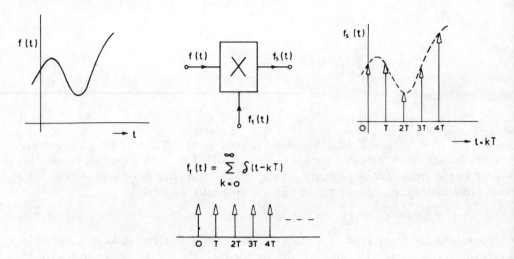

Fig. 1.7   Modulation of a signal by a sequence of impulses

Therefore, the Laplace transform of the sampled waveform described by equation 1.16 is given by

$$F_s(s) = \sum_{k=0}^{\infty} f(kT)e^{-skT} \tag{1.17}$$

Comparing equation 1.17 with equation 1.9 and taking into account the change in notation discussed above, we have

$$F(z) = F_s(s)\Big|_{e^{sT} = z} \tag{1.18}$$

which establishes the relationship between z and complex frequency s, i.e.

$$z = e^{sT} \tag{1.19}$$

This is an important relationship and it is examined in some detail in section 2 of the appendix.

There may be differences in the treatment of equation 1.16 and those following from it. For example, Kaiser(9) multiplies equation 1.16 by T and defines the z-transform as $TF(z)$, where $F(z)$ is given by equation 1.9. The factor T is explained as an approximation of $f_s(t)$ to $f(t)$ in the sense that the area under both functions is approximately the same in the interval $kT<t<(k+1)T$, provided that T is sufficiently small. It seems to be more correct to include T in equation 1.16 and other related expressions, since a multiplier such as the cutoff frequency ω_c then becomes $\omega_c T$ which is dimensionally correct (see section 4.1). We shall keep equation 1.17 as shown since we have defined $F(z)$ by equation 1.9, which is the commonly-used form.

The Laplace transform, equation 1.17, enabled us to link s-plane with z-plane as expressed by equation 1.19. The importance of this relationship will be seen in the following chapters. There is another form of Laplace transform of the sampled wave (section 3 of the appendix) given by

$$F_s(s) = \frac{1}{T} \sum_{n=-\infty}^{\infty} F(s + jn\omega_s) \qquad (1.20)$$

where $\omega_s = 2\pi/T$ represents the sampling frequency (rad s^{-1}), and $F(s)$ is the Laplace transform of the input signal. The roots of $F_s(s)$ are periodic in the s-plane as shown in fig. 1.8(a), and so is $F_s(s)$ as shown in figs 1.8(b) and (c). The main point of fig. 1.8(b),

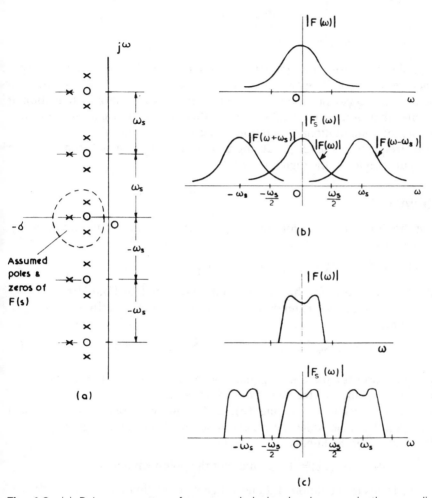

Fig. 1.8 (a) Pole zero pattern for a sampled signal, where ω_s is the sampling frequency; (b) frequency spectra before sampling $|F(\omega)|$, and after sampling $|F_s(\omega)|$; (c) $|F(\omega)|$ and $|F_s(\omega)|$, but $|F(\omega)|$ is well bandlimited to frequencies below $\omega_s/2$. Note that this spectra shows continuous functions of frequency

also expressed in fig 1.8(c), is that in order to avoid interspectra interference (or frequency aliasing), the input signal must be bandlimited to frequencies below $\omega_s/2$. As will be seen later, the digital filter design is confined to the frequency range $-\omega_s/2 \leq \omega \leq \omega_s/2$. The required filter characteristic (lowpass, bandpass, etc.) is then specified within this frequency range.

The general scheme of continuous-signal simulation in discrete-time form is shown in

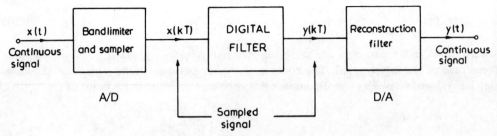

Fig. 1.9   Continuous-signal processing in discrete-time form

fig. 1.9. The bandlimiter before the sampler is a good lowpass filter to restrict the input signal to frequencies below $\omega_s/2$. Such a filter should have a flat magnitude response and a linear phase over the major portion of the interval $\pm\omega_s/2$, and a large attenuation at frequencies greater than $\omega_s/2$. A minimal sampling rate is usually best since it results in a minimum number of computational operations. The reconstruction filter is a lowpass filter with cutoff frequency of $\omega_s/2$, producing a continuous-time signal from the time sequence of output samples of the digital filter.

1.5   Problems

1.1   Show that the difference equation for the highpass connection of the circuit in fig. 1.1(a) is given by

$$y(k) = (1 - \Delta t/RC)[y(k-1) + x(k) - x(k-1)]$$

1.2   (a) Write the z-transform for the finite sequence given by $[1,0,1/4,0,1/16]$.
(b) If the above sequence extends to infinity, i.e. $[1,0,1/4,0,1/16,0,1/64,0,\ldots]$, write its z-transform.

1.3   (a) Plot the exponential signal $x(k) = c^k$ over the range $k = 0, 1, 2, \ldots, 10$, for $c = \pm0.5, \pm1.5$.
(b) The z-transform for the above signal is given in Table 1.1. Plot its pole-zero positions in the z-plane, for the values of c given in part (a).
(c) Compare the results of (a) and (b), and show why some cases represent bounded (stable) signals while the others unbounded (unstable) signals; see, for example, section 2 of the appendix.

1.4   Find the z-transform of $x(k) = 1 + k$, and plot the pole-zero pattern.

1.5   Write z-transforms for equations 1.2 and 1.4 given in section 1.1.

1.6   Apply z-transform to the set of equations 1.8 and show that

$$Y(z) = \frac{z(\alpha z + \beta - \alpha)}{D} X(z)$$

$$\dot{Y}(z) = \frac{\beta}{T} \frac{z(z-1)}{D} X(z)$$

$$Y_p(z) = \frac{z(\alpha + \beta) - \alpha}{D} X(z)$$

where $D = z^2 + (\beta + \alpha - 2)z + (1 - \alpha)$.

1.7 Assuming that in equations 1.8 $x(k) = \delta(k)$ is a unit pulse, show that, for the equations in problem 1.6 to represent a critically damped (nonoscillatory) system, the relationship between α and β must satisfy

$$\alpha = 2\sqrt{\beta} - \beta.$$

1.8 Show that the final value $f(\infty)$ is obtained from

$$f(\infty) = \operatorname*{Lim}_{z \to 1} (z-1)F(z) \tag{1.21}$$

Note: This only applies for cases in which $(z-1)F(z)$ is analytic for $|z| \geq 1$ (see, for example, Cadzow(4)).

Hint: Start with

$$Z[f(k+1) - f(k)] = \operatorname*{Lim}_{N \to \infty} \sum_{k=0}^{N} [f(k+1) - f(k)]z^{-k}$$

1.9 (a) Apply z-transform to equation 1.6 and show that

$$Y(z) = H(z)X(z)$$

where $\quad H(z) = \sum_{i=0}^{k} h(i)z^{-i}$

(b) Let x and h be finite length sequences each consiting of two terms: $x = (2, 1)$ and $h = (3,4)$. Determine their convolution sequence firstly from $y = h * x$ (see equation 1.6), and secondly from $Y(z) = H(z)X(z)$.

2
Digital filter characterization

2.0 Introduction

The basic analytical tools and relationships which have been introduced so far enable us to develop the concepts similar to analogue filters. In the first section of this chapter, we derive the digital filter transfer function $H(z)$ in the z-domain, and explore its properties by calculating the impulse response and frequency response. Several simple examples are used to illustrate methods of analysis.

We also discuss the digital filter hardware structures and consider their relative merits. The last section deals with the classification of digital filters into nonrecursive and recursive types, and the general properties of these two classes are briefly discussed.

2.1 Digital transfer function

We have already seen that in a linear discrete-time system, the input $x(k)$ and output $y(k)$ sequences are related by linear difference equations with constant coefficients as in equations 1.2 and 1.4. In the theory of digital or discrete-time filters, the general difference equation is usually written in the following way:

$$y(k) + b_1 y(k-1) + \ldots + b_N y(k-N)$$
$$= a_0 x(k) + a_1 x(k-1) + \ldots + a_M x(k-M) \tag{2.1}$$

where b_0 is taken, by convention, as unity.

The interpretation of equation 2.1 is that at time $k(t = kT)$, the output value can be computed from the current input and a linear combination of previous inputs and outputs. The output sequence, or response of a digital filter for a given input sequence, can therefore be calculated in a simple manner as illustrated below in example 2.1.

Example 2.1
Taking $a_0 = 1$, $a_i = 0$ for $i \neq 0$, and $b_1 = \pm 0.8$, $b_i = 0$ for $i \geq 2$, equation 2.1 reduces to

$$y(k) = x(k) \mp 0.8 y(k-1) \tag{2.2}$$

If the input is a unit pulse described as $x(0) = 1$ and zero for all $k \neq 0$, and assuming initial conditions to be $x(k) = y(k) = 0$ for $k < 0$, we can form the calculation scheme shown in Table 2.1. The signs $\mp$ refer to filters with $b_1 = \pm 0.8$. The input and output sequences for this case are shown in fig. 2.1, where the full lines are for $b_1 + 0.8$ and also dashed ones for $b_1 = -0.8$.

If the input is a unit step given by $x(k) = 1$ for $k \geq 0$, with the same initial conditions as

16

Table 2.1

k	$x(k)$	$y(k)$	$y(k-1)$
0	1	1	0
1	0	$\mp 0{\cdot}80$	1
2	0	0·64	$\mp 0{\cdot}80$
3	0	$\mp 0{\cdot}51$	0·64
4	0	0·41	$\mp 0{\cdot}51$
5	0	$\mp 0{\cdot}33$	0·41
6	0	0·26	$\mp 0{\cdot}33$

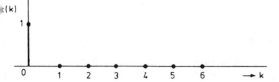

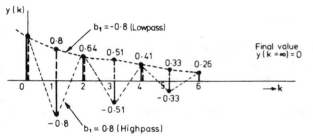

Fig. 2.1 Filter response for the unit pulse input

Table 2.2

		$b_1 = 0{\cdot}8$		$b_1 = -0{\cdot}8$	
k	$x(k)$	$y(k)$	$y(k-1)$	$y(k)$	$y(k-1)$
0	1	1	0	1	0
1	1	0·20	1	1·8	1
2	1	0·84	0·20	2·44	1·8
3	1	0·33	0·84	2·95	2·44
4	1	0·74	0·33	3·35	2·95
5	1	0·41	0·74	3·69	3·35
6	1	0·67	0·41	3.95	3·69

in the above case, we obtain Table 2.2. The input and output sequences for this case are shown in fig. 2.2. the final values $y(\infty)$ are also given in figs 2.1 and 2.2. They have been calculated using the final value equation 1.21 established in problem 1.8; $y(\infty)$ is a useful quantity since it shows whether the output sequence converges. We shall refer to figs 2.1 and 2.2 again in example 2.4.

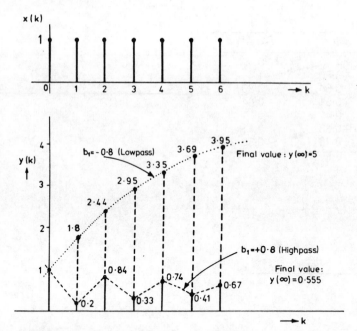

Fig. 2.2   Filter response for the unit step input

The above illustrated step-by-step solution of the difference equation for a given input is a straightforward process, but if we require a closed-form solution and frequency response of the output sequence, we need an alternative approach based on the z-transform.

Taking the z-transform of equation 2.1 term-by-term, and using equations 1.9 and 1.14 as appropriate, we obtain

$$Y(z)\left(1 + \sum_{i=1}^{N} b_i z^{-i}\right) = X(z) \sum_{i=0}^{M} a_i z^{-i} \tag{2.3}$$

where $X(z) = \sum_{k=0}^{\infty} x(k)z^{-k}$ and $Y(z) = \sum_{k=0}^{\infty} y(k)z^{-k}$

From equation 2.3 we can now define the discrete-time (or digital) transfer function as

$$H(z) = \frac{Y(z)}{X(z)} = \frac{\sum_{i=0}^{M} a_i z^{-i}}{1 + \sum_{i=1}^{N} b_i z^{-i}} \tag{2.4}$$

which is a rational polynomial in z^{-1}. This transfer function is valid for zero initial conditions which are satisfied in the cases we are considering. For nonzero initial conditions, see, for example, Cadzow(4), p. 227.

We have seen that the z-transformation of the difference equation 2.1 enabled us to define the digital transfer function. This is analogous to the Laplace transform of a differential equation in continuous-time, which produces the transfer function $H(s)$ in s-domain. From equation 2.4 we can write for the output

$$Y(z) = H(z)X(z) \tag{2.5}$$

and the output sequence $y(k)$ is then obtained using the inverse z-transform. The special case of equation 2.5 is obtained for the unit pulse input sequence

$$x(k) = \begin{cases} 1 & k = 0 \\ 0 & k \neq 0 \end{cases}$$

which has z-transform $X(z) = 1$. The response to this input is then the inverse z-transform of $H(z)$. Therefore, the input sequence $[1, 0, 0, \ldots]$ in discrete-time filter theory corresponds to the unit impulse in continuous-time filter theory, and the inverse z-transform of $H(z)$ is the impulse response of a digital filter. We note here that the impulse function in discrete-time, as defined above, is a more realistic quantity than the corresponding impulse function in continuous-time.

2.2   The inverse transformation

To obtain the impulse response sequence $h(k)$ from $H(z)$, or $y(k)$ from $Y(z)$, we can use one of the three methods of inversion:

(i)   inversion by series expansion (long division);
(ii)   inversion by partial-fraction expansion;
(iii)   use of an inversion integral described by Freeman(6) and Oppenhein & Schaffer(10).

We illustrate the first two methods of inversion using the filter of example 2.1.

Example 2.2
Applying z-transform to equation 2.2 we obtain the following transfer function:

$$H(z) = \frac{1}{1 \pm 0.8z^{-1}} = \frac{z}{z \pm 0.8} \tag{2.6}$$

The inversion of $H(z)$, by direct division of the numerator of the above equation by the denominator using long division, produces the series

$$H(z) = 1 \mp 0.8z^{-1} + 0.64z^{-2} \mp 0.51z^{-3} + 0.41z^{-4} \mp 0.33z^{-5} + 0.26z^{-6} + \ldots$$

The coefficients in this expansion are the values of the impulse response: $h(0) = 1$, $h(1) = \mp 0.8$, $h(2) = 0.64$, $h(3) = \mp 0.51$, etc. We note that these coefficients are the same as $y(k)$ values in Table 2.1.

The series expansion is useful when the first few terms of the sequence $h(k)$ are to be found. The general solution in closed form, is obtained using partial-fraction expansion. The inversion of $H(z)$ by partial-fractions, in this case, is obtained by recognizing that $H(z)$ in equation 2.6 is the third entry in Table 1.1. We then have immediately $h(k) = \mp 0.8^k$, which is the closed form solution of the above obtained impulse response series.

Example 2.3
The unit step input has the z-transform

$$X(z) = \frac{1}{1 - z^{-1}} \tag{2.7}$$

given by the second entry in Table 1.1. The z-transformed output is then obtained from equation 2.5 for the filter, with $b_1 = 0.8$, specified by equation 2.6, as

$$Y(z) = \frac{z^2}{z^2 - 0.2z - 0.8} \tag{2.8}$$

Dividing the numerator by the denominator we obtain the following series:

$$Y(z) = 1 + 0.2z^{-1} + 0.84z^{-2} + 0.33z^{-3} + 0.738z^{-4} + 0.412z^{-5} + 0.672z^{-6} + \ldots$$

whose coefficients represent the filter output sequence for the unit step input. Note that these values are the same as in Table 2.2.

For the partial-fraction method we have to determine the poles which, for equation 2.8, are found to be 1 and -0.8. Then we rewrite equation 2.8 as

$$Y(z) = z\left[\frac{z}{(z-1)(z+0.8)}\right] \tag{2.9}$$

The function inside the square bracket has the degree of its numerator less than the degree of its denominator. Such rational functions are called proper rational functions, and they can be represented as a sum of simple one-pole terms as follows:

$$\frac{z}{(z-1)(z+0.8)} = \frac{A}{z-1} + \frac{B}{z+0.8} \tag{2.10}$$

The constants A and B can be found in the following manner. Multiply both sides of equation 2.10 by $(z-1)$, and setting $z = 1$ we have

$$A = \left.\frac{z}{z+0.8}\right|_{z=1} = \frac{1}{1.8}$$

Now multiplying both sides of equation 2.10 by $z + 0.8$, and setting $z = -0.8$, we obtain

$$B = \left.\frac{z}{z-1}\right|_{z=-0.8} = \frac{1}{2.25}$$

Therefore, equation 2.9 can be written as

$$Y(z) = \frac{1}{1.8}\left(\frac{z}{z-1}\right) + \frac{1}{2.25}\left(\frac{z}{z+0.8}\right)$$

or

$$Y(z) = \frac{1}{1.8}\left(\frac{1}{1-z^{-1}}\right) + \frac{1}{2.25}\left(\frac{1}{1+0.8z^{-1}}\right) \tag{2.11}$$

The factors in brackets are the second and third entries in Table 1.1, from which we obtain corresponding time sequences resulting in the output sequence

$$y(k) = \frac{1}{1.8} + \frac{1}{2.25}(-0.8)^k \tag{2.12}$$

This is the closed form solution which can be checked for $k = 0, 1, 2, 3, \ldots$ with the coefficients obtained in the earlier series expansion.

In the above example poles of $Y(z)$ have been simple-order poles. A similar method applies to multiple-pole cases, and is described by Cadzow(4) and Proakis & Manolakis(11).

2.3 Frequency response

With regard to filtering we are particularly interested in the interpretation of the transfer (or system) function $H(z)$ as a frequency selective function. Therefore, we consider difference equation 2.1 when the input is a sampled exponential waveform

$$x(k) = e^{jk\omega T} \quad \text{for } k = 0, 1, 2, 3, \ldots \tag{2.13}$$

We can take then that the output is

$$y(k) = F(\omega) e^{jk\omega T} \tag{2.14}$$

since the system is linear, and hence the output is of the same frequency as the input, with $F(\omega)$ being a complex proportionality factor depending on ω only and determined as follows.

Using equation 2.13 and 2.14 in equation 2.1, we have

$$e^{jk\omega T} F(e^{j\omega T})(1 + b_1 e^{-j\omega T} + b_2 e^{-j2\omega T} + \ldots + b_N e^{-jN\omega T})$$
$$= e^{jk\omega T}(a_0 + a_1 e^{-j\omega T} + a_2 e^{-j2\omega T} + \ldots + a_M e^{-jM\omega T})$$

and

$$F(e^{j\omega T}) = \frac{\sum_{i=0}^{M} a_i e^{-ji\omega T}}{1 + \sum_{i=1}^{N} b_i e^{-ji\omega T}} \tag{2.15}$$

Comparing equation 2.15 with equation 2.4, we see that

$$F(e^{j\omega T}) = H(z)\Big|_{z = e^{j\omega T}} = H(e^{j\omega T}) \tag{2.16}$$

Thus the frequency response of a discrete-time filter is determined by the values of its transfer function on the unit circle in the z-plane given by $|z| = |e^{j\omega T}| = 1$. This is again analogous to the frequency response of a continuous-time filter being determined by the values of its transfer function on the imaginary axis. The s–z plane correspondence is examined in more detail in section 2 of the appendix.

Example 2.4
To determine the frequency response of the filter described by the transfer function 2.6, we set $z = e^{j\omega T}$ and obtain

$$H(\omega) = \frac{1}{1 \pm 0.8 e^{-j\omega T}}$$

Expanding $e^{-j\omega T}$ in the usual manner, and calculating the magnitude response we have

$$|H(\omega)| = \frac{1}{1.28\sqrt{(1 \pm 0.975 \cos \omega T)}}$$

Evaluating this function at points $\omega T = 0, \pi/4, \pi/2, 3\pi/4, \pi$, we obtain the result plotted in fig. 2.3. The abscissa is marked in angle ωT and frequency ω. It is seen that $|H(\omega)|$ is a periodic and continuous function of ω. However, it is sufficient to specify the filter

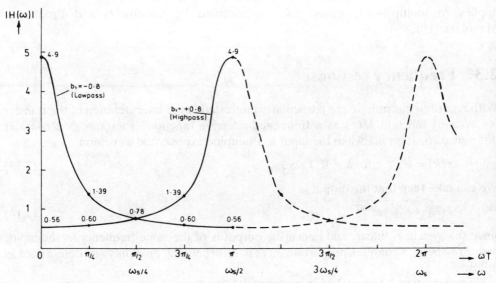

Fig. 2.3   Frequency response of the filter specified by equation 2.6

characteristics in the frequency range $0 \leq \omega \leq \omega_s/2$, i.e. up to half of the sampling frequency ($\omega_s/2 = \pi/T$) as discussed in section 1.4, and particularly fig. 1.8(c).

The graph for $b_1 = +0.8$ shows that the filter is a highpass in the frequency range $0 \leq \omega \leq \omega_s/2$, but for $b_1 = -0.8$, we have a lowpass filter in the same frequency range. Therefore, as expected, the frequency selective properties of the filter are clearly shown in the frequency-domain. These properties are not as clearly shown in the time-domain graphs in figs 2.1 and 2.2, but even there, one can observe that the output of the filter with $b_1 = -0.8$ allows slow time variations (lowpass filter in figs 2.1 and 2.2), while the output of the filter with $b_1 = 0.8$ allows fast time variations to pass through (highpass filter in figs 2.1 and 2.2).

It is perhaps a suitable place here to point out that discrete-time processing of signals is much simpler to deal with than continuous-time processing, since finding the response of a difference equation to inputs such as impulse or step is a straightforward algebraic calculation which can be done in a short time. However, the corresponding time-domain analysis of analogue filters means that we have to solve a differential equation every time we change its order or the input signal.

2.4   Computation of the frequency response

As shown in the previous section, the frequency response can be obtained from equation 2.4 by setting $z = e^{j\omega T}$. Computation of the magnitude and phase of such a function would require the use of a suitable computer program, unless $H(z)$ is a simple first order function. An interesting alternative is to use a geometric (graphic) approach in the z-plane.

To explain this we start with equation 2.4 which is a ratio of two polynomials. We denote roots of the numerator polynomial by z_i (they are zeros of $H(z)$), and we denote roots of the denominator polynomial by p_i (representing poles of $H(z)$). Then, to simplify we take, in equation 2.4, $M = N$, and write $H(z)$ in terms of poles and zeros as

$$H(z) = G\frac{\prod_{i=1}^{N}(z-z_i)}{\prod_{i=1}^{N}(z-p_i)} \tag{2.17}$$

where G is a constant gain factor.

The frequency response is obtained by setting $z = e^{j\omega T}$, so that the magnitude response is given by

$$|H(\omega)| = G\frac{\prod_{i=1}^{N}|(e^{j\omega T}-z_i)|}{\prod_{i=1}^{N}|(e^{j\omega T}-p_i)|} \tag{2.18}$$

and the phase response

$$\arg H(\omega) = \sum_{i=1}^{N}[\arg(e^{j\omega T}-z_i) - \arg(e^{j\omega T}-p_i)] \tag{2.19}$$

The quantities $(e^{j\omega T}-z_i)$ and $(e^{j\omega T}-p_i)$ are vectors from z_i and p_i to the point $e^{j\omega T}$ on the unit circle in the z-plane. Each pole and zero contributes an independent multiplication term for the magnitude, and an additive term for the phase. These contributions can be obtained graphically using the pole-zero diagram as shown in the following examples.

Example 2.5
The transfer function of a second-order system is given as

$$H(z) = \frac{2-3z^{-1}+z^{-2}}{1-z^{-1}+0.5z^{-2}} = 2\frac{z^2-1.5z+0.5}{z^2-z+0.5}$$

Therefore $G = 2$, zeros are at $z_1 = 1$, $z_2 = 0.5$, and poles at $p_{1,2} = 0.5(1\pm j)$. Their positions are indicated in fig. 2.4 by dashed vectors. The frequency point is given by the

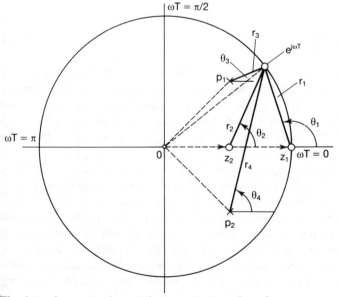

Fig. 2.4 Geometric (graphic) determination of the frequency response (magnitude and phase)

unity vector $e^{j\omega_1 T}$ for some chosen frequency ω_1. The vector differences for equation 2.18 are:

$$r_1 = |e^{j\omega_1 T} - z_1|, r_2 = |e^{j\omega_1 T} - z_2|, r_3 = |e^{j\omega_1 T} - p_1|, r_4 = |e^{j\omega_1 T} - p_2|$$

The angles for equation 2.19 are denoted by $\theta_1, \theta_2, \theta_3, \theta_4$. The magnitude and phase at frequency ω_1 are given by

$$|H(\omega_1) = 2 \cdot \frac{r_1 r_2}{r_3 r_4}$$

$$\arg H(\omega) = (\theta_1 + \theta_2) - (\theta_3 + \theta_4) \tag{2.20}$$

Varying the frequency between $\omega T = 0$ up to $\omega T = \pi$, we can compute the frequency response point by point using equation 2.20.

The computation is straightforward, but the results would be only approximate due to the nature of the graphical approach. However, this representation also shows some interesting and useful features. For example if a frequency ω_0 is to be rejected we simply place a pair of zeros $e^{\pm j\omega_0 T}$ on the unit circle. If the frequency ω_0 is to be emphasized, place a pair of poles inside the circle near $e^{\pm j\omega_0 T}$.

2.5   Digital filter realization schemes

Digital filter realization of a given transfer function $H(z)$ is quite simple. Suppose the input is $x(k)$ with z-transform $X(z)$, and the output is $y(k)$ with z-transform $Y(z)$. Then, from equation 2.4, we have

$$Y(z) = \sum_{i=0}^{M} a_i z^{-i} X(z) - \sum_{i=1}^{N} b_i z^{-i} Y(z)$$

and applying equations 1.9 and 1.14 in inverse sense, we obtain the time-domain equation

$$y(k) = \sum_{i=0}^{M} a_i x(k-i) - \sum_{i=1}^{N} b_i y(k-i) \tag{2.21}$$

which is the difference equation that realizes $H(z)$ directly, and it also represents a computational algorithm. A hardware implementation of equation 2.21 is shown in fig. 2.5. As before, in figs 1.2(a) and 1.4, a triangle represents multiplication of a variable by the constant written beside it, a rectangle represents a unit-sample delay, and a circle represents a summing point. The interpretation of fig.2.5 is as follows: at $t = kT$, $x(k)$ becomes available and the quantities $x(k-1), x(k-2), \ldots, x(k-M), y(k-1), \ldots, y(k-N)$ at the outputs of the delay elements have been remembered, i.e. stored. Thus all the variables are available for the computation of $y(k)$. When this computation is complete $x(k-M)$ and $y(k-N)$ are discarded, but the other quantities are saved since they will be needed for the next computation, i.e. input $x(k+1)$ and outputs $y(k+1)$. In this way, an entire input sequence of infinite duration can be filtered by the algorithm 2.21 to produce an output sequence of the same length. By counting the number of delays in fig. 2.5, we obtain a minimum number of storages required in realizing equation 2.21, and by counting the number of triangles we can see how many multiplications are required per sample.

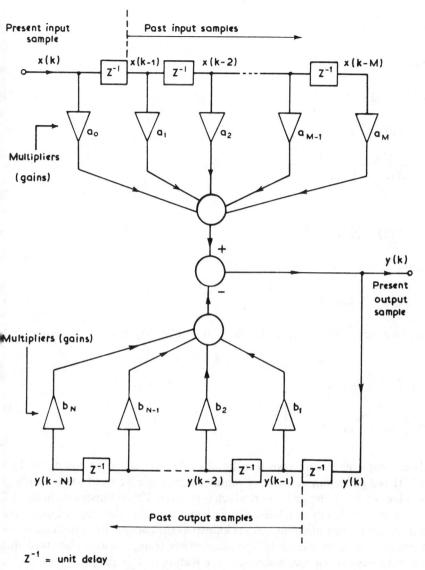

Present input
sample

$x(k)$

Past input samples

$x(k-1)$ $x(k-2)$ $\cdots$ $x(k-M)$

Z^{-1} Z^{-1} Z^{-1}

a_0 a_1 a_2 a_{M-1} a_M

Multipliers

(gains)

$+$

$y(k)$

Present
output
sample

$-$

Multipliers (gains)

b_N b_{N-1} b_2 b_1

Z^{-1} Z^{-1} Z^{-1}

$y(k-N)$ $y(k-2)$ $y(k-1)$ $y(k)$

Past output samples

Z^{-1} = unit delay

Fig. 2.5   Hardware implementation of a discrete-time filter

Equation 2.21 and the corresponding fig. 2.5 is not the only possible way to realize a given digital filter function $H(z)$. For example, we can introduce an intermediate variable $W(z)$ by partitioning equation 2.4 in the following way:

$$H(z) = \frac{Y(z)}{X(z)} = \frac{Y(z)}{W(z)} \cdot \frac{W(z)}{X(z)} = N(z)\frac{1}{D(z)}$$

The first part is

$$N(z) = \frac{Y(z)}{W(z)} = \sum_{i=0}^{M} a_i z^{-i}$$

leading to $Y(z) = \sum_{i=0}^{M} a_i z^{-i} W(z)$

or in time-domain

$$y(k) = \sum_{i=0}^{M} a_i w(k-i) \tag{2.22}$$

Similarly the second part is

$$\frac{1}{D(z)} = \frac{W(z)}{X(z)} = \frac{1}{1 + \sum_{i=1}^{N} b_i z^{-i}}$$

leading to

$$W(z) = X(z) - \sum_{i=1}^{N} b_i z^{-i} W(z)$$

or in time-domain

$$w(k) = x(k) - \sum_{i=1}^{N} b_i w(k-i) \tag{2.23}$$

Therefore, in place of equation 2.21 we have the following set of computational algorithms:

$$\left.\begin{aligned} w(k) &= x(k) - \sum_{i=1}^{N} b_i w(k-i) \\ y(k) &= \sum_{i=0}^{M} a_i w k - i) \end{aligned}\right\} \tag{2.24}$$

with the hardware implementation shown in fig. 2.6. An advantage of equations 2.24 over equation 2.21 is a reduction in memory requirement, since we need to save only N or M previous values of $w(k)$, depending on which is greater. This is illustrated in fig. 2.7 (redrawn fig. 2.6 with $N = M$) in which the delay elements having the same output have been replaced by a single delay element. This is known as the canonic form because it has the minimum number of delay elements, but since other configurations also have this property, this terminology is not recommended (see Rabiner(1), p. 328).

The realizations of digital filters shown in figs 2.5 and 2.7 are called direct forms. It is interesting to note that the constants a_n and b_m in the network, i.e. hardware implementation, are the same as the constants in the transfer function. In continuous-time systems this is not the case since the network elements (R, L, C) are not so easily related to the transfer function of a continuous-time filter.

The direct form realization of $H(z)$ in equation 2.4 is attractive for its simplicity, but undesirable for high-order systems because of their high sensitivity to coefficient variations. It is then necessary to express $H(z)$ by equation 2.17 and form second-order factors leading to cascade (or serial) structure, fig. 2.8, for which

$$H(z) = \prod_{k=1}^{K} H_k(z) \tag{2.25a}$$

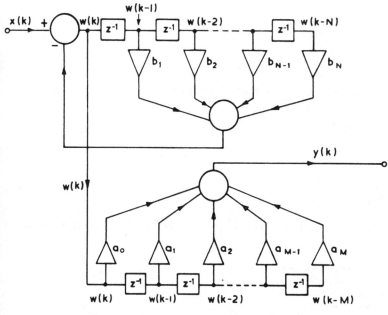

Fig. 2.6   Alternative hardware implementation of discrete-time filter

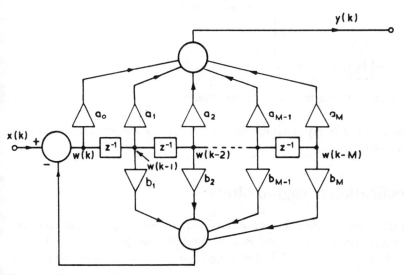

Fig. 2.7   Representation of fig. 2.5 with a reduced number of delay elements (canonic form), for $N = M$

Fig. 2.8   Serial connection for first- and second-order subfilters

where

$$H_k(z) = \frac{a_{k0} + a_{k1}z^{-1} + a_{k2}z^{-2}}{1 + b_{k1}z^{-1} + b_{k2}z^{-2}} \qquad (2.25b)$$

For the first order sections $a_{k2} = b_{k2} = 0$.

Alternatively expanding equation 2.4 in partial fractions (with $M \geq N$) we have

$$H(z) = C + \sum_{k=1}^{N} H_k(z) \qquad (2.26a)$$

where

$$H_k(z) = \frac{a_{k0} + a_{k1}z^{-1}}{1 + b_{k1}z^{-1} + b_{k2}z^{-2}} \qquad (2.26b)$$

which is illustrated in fig 2.9. When N is odd, one of $H_k(z)$ is a first-order section

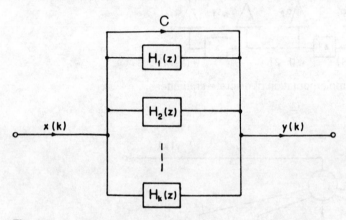

Fig. 2.9 Parallel connection of first- and second-order subfilters

$(a_{k1} = b_{k2} = 0)$. More detail is given about the above decomposition, for example in Proakis and Manolakis(11).

2.6   The classification of digital filters

For the purpose of realization, digital filters are classified into nonrecursive and recursive types. The nonrecursive structure contains only the feed-forward paths as shown in fig. 2.10. This is a special case of equation 2.4 in which all b_i coefficients are zero, so that the transfer function is given by

$$H(z) = \sum_{i=0}^{M} a_i z^{-i} \qquad (2.27)$$

In recursive filter structures the output depends both on the input and on the previous outputs, as shown in the general hardware realization of fig 2.5, where we have both feed-forward and feed-back paths.

An alternative division of digital filters is made, on the basis of the impulse response

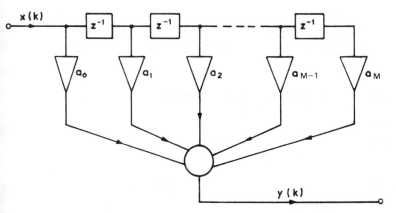

x (k)

a_0 a_1 a_2 a_{M-1} a_M

y (k)

Fig. 2.10   Nonrecursive discrete-time filter

duration, into finite response (FIR) and inifinite response (IIR) filters. The simplest FIR filter realization is in the nonrecursive form. For example, a nonrecursive filter

$$y(k) = x(k) + a_1 x(k-1) \tag{2.28}$$

has a finite impulse response given by $[1, a_1, 0, 0, \ldots]$, but the simplest IIR filter realization is in the recursive form, for example,

$$y(k) = x(k) - b_1 y(k-1) \tag{2.29}$$

with impulse response $[1, -b_1, b_1^2, -b_1^3, b_1^4, \ldots]$, which has an infinite number of terms.

The division based on the length of the unit impulse response has been generally accepted. Therefore, in the following we shall use FIR for the nonrecursive filters and IIR for the recursive filters.

It is of interest to note also that in time series analysis the nonrecursive filter is known as the moving-average filter, and the recursive filter is the autoregressive moving-average filter, as discussed by Anderson(5).

2.7   Problems

2.1   For a filter described by the linear difference equation

$$y(k) = x(k) + 0.5y(k-1)$$

determine the first five terms of its reponse to the input

$$x(k) = \begin{cases} 0 & k < 0 \\ 1, -2, 2, 1, -2 & k = 0, 1, 2, 3, 4, \text{ respectively.} \end{cases}$$

Consider two different initial conditions: (a) $y(-1) = 0$ and (b) $y(-1) = 2$. Solve this problem in time-domain.

2.2   Solve the problem specified above in problem 2.1 using z-transform. Compare the result obtained with the solution of problem 2.1 for $y(-1) = 0$.

2.3 Solve the problem specified in problem 2.1, with the initial condition $y(-1) = 2$, using z-transform. Compare the result with the solution in problem 2.1.
Note: In this case use modified input z-transform given by

$$\overline{X}(z) = X(z) + 0.5y(-1)$$

2.4 Derive a general expression for the response of the filter

$$y(z) = a_0 x(k) - b_1 y(k-1)$$

to the input signal

$$x(k) = \begin{cases} 0 & k < 0 \\ 1 & k = 0, 2, 4, 6, \ldots \\ -1 & k = 1, 3, 5, 7, \ldots \end{cases}$$

The initial condition is $y(-1) = 0$. Develop the solution by working in time-domain.

2.5 Determine the first five terms of the inverse transform of the following two functions:

$$X_1(z) = \frac{z^{-1} + z^{-2}}{1 - z^{-2}}, \quad X_2(z) = \frac{1 + 2z^{-1}}{1 - \frac{1}{2}z^{-1}}$$

(a) by long division and (b) by partial-fraction expansion.

2.6 Determine the first five terms of the inverse transform of the function

$$X(z) = \frac{1}{1 - z^{-1} + \frac{1}{2}z^{-2}}$$

(a) by long-division and (b) by partial-fraction expansion.

2.7 (a) Determine the difference equation relating $x(k)$ and $y(k)$ for the system shown in fig. 2.11. Write its transfer function $H(z)$.

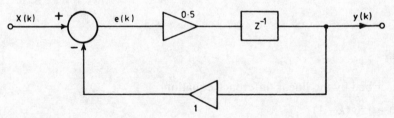

Fig. 2.11

(b) Using the transfer function, determine the first six terms of the impulse response of this system.
(c) Derive the amplitude and phase response of this system. What kind of filtering does it perform?

2.8 Examine the effect on the filter

$$y(k) = \frac{1}{2}x(k) + \frac{1}{2}x(k-1)$$

if, in its transfer function, z is replaced by z^2. More specifically, how are the amplitude and phase response affected? Plot simple graphs of these quantities as functions of ωT.

2.9 In radar systems a moving target indicator employs a filter to remove the background clutter, as described fully by Skolnik(12). The simplest type of such a filter is the single delay line canceller shown in fig. 2.12 as Unit 1. Calculate the

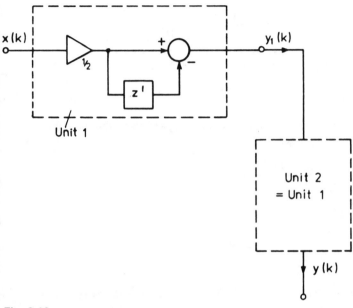

Fig. 2.12

amplitude and phase response of this unit (output y_1), and also for the cascade of two such units (output y). Compare the results by sketching simple graphs at points $\omega T = 0$, $\pi/2$, π.

2.10 Calculate the output sequence of the double delay canceller (cascade of the two units in fig. 2.12 of problem 2.9), for:
(a) unit input step, $x(k) = 1$ for $k \geq 0$;
(b) ramp input, $x(k) = k$ for $k \geq 0$.
In both cases $x(k) = 0$ for $k < 0$.
 Perform a step-by-step analysis in the time-domain, and also using the z-transform method, $Y(z) = H(z)X(z)$.
Note: This case is also known as the three-pulse canceller. The reason for this name will be found in the course of the above specified analysis; work up to at least $k = 5$.

2.11 (a) Determine the difference equation relating $x(k)$ and $y(k)$ for the system shown in fig. 2.13.
(b) Derive the frequency response and calculate the amplitude response at $\omega T = 0$, $\pi/2$ and π. Sketch the graph passing through these points and compare it with the graphs for the amplitude response of the canceller in problem 2.9.

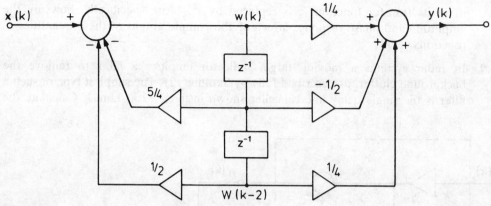

Fig. 2.13

2.12 The system of equations governing α–β tracker, as given in problems 1.6 and 1.7, can be written in the following way:

$$H_1(z) = \frac{z[(2\sqrt{\beta}-\beta)z+2\beta-2\sqrt{\beta}]}{(z+\sqrt{\beta}-1)^2}$$

$$H_2(z) = \frac{\beta}{T}\frac{z(z-1)}{(z+\sqrt{\beta}-1)^2}$$

$$H_3(z) = \frac{(2\sqrt{\beta})z+\beta-2\sqrt{\beta}}{(z+\sqrt{\beta}-1)^2}$$

where $H_1 = Y/X_0$, $H_2 = \dot{Y}/X$ and $H_3 = Y_p/X$ are the three transfer functions.
(a) Show that for the tracker to have stable dynamics, β must be within the range $0<\beta<4$.
(b) Calculate and compare graphically the impulse responses of $H_1(z)$ for $\beta = 0.81$ and for $\beta = 3.6$.

2.13 The transfer function of a discrete-time filter is given by

$$H(z) = \frac{5z^2 - 12z}{z^2 - 6z + 8}$$

(a) Show that the first four values of its impulse response are $h(0) = 5$, $h(1) = 18$, $h(2) = 68$, $h(3) = 264$.
(b) Show also that the closed form impulse response is given by

$$h(k) = 2^k + 4^{k+1}$$

2.14 Determine the transfer function of an FIR filter such that it rejects a 50 Hz sinusoidal interference, but it passes the useful signal at 150 Hz without changing its amplitude. The sampling frequency is $f_s = 500$ Hz.
Note: use equation 2.17 with $N = 2$ and $p_1 = p_2 = 0$.

3

FIR filter design

3.0   Introduction

In the previous chapter we have further developed the analytical tools which enabled us to perform the time and frequency analysis of given digital filters, and we have classified filters and discussed their properties. The aim of this chapter and the next is to use the techniques developed so far and to produce systematic procedures for the design of digital filters.

The design methods for each of these two classes of filters are different because of their distinctly different properties. The FIR filter has a finite memory and can have excellent linear phase characteristics, but it requires a large number of terms, possibly in the order of several hundred, to obtain a relatively sharp cutoff frequency response. However, program execution time for FIR filters can be considerably reduced using the Fast Fourier Transform techniques. The IIR filter has an infinite memory and tends to have fewer terms, but its phase characteristics are not as linear as FIR. IIR digital filters usually meet the stringent specifications arising in practice with at most ten to twenty coefficients, so the computation required to produce each output, given a new input, is of the order of ten to twenty multiplications and additions per sample point.

Filter design can be implemented in the time-domain or frequency-domain. We shall be dealing with the design of filters specified in the frequency-domain because it is more precise and more suitable for electrical engineering practice. The time-domain approach may be useful to give an initial guess as described by Rabiner & Gold(13), p. 265.

The filter frequency characteristic is specified in so-called baseband range $0 \le \omega \le \omega_s/2$ where ω_s is the sampling frequency. Typical idealized lowpass, highpass, bandpass and bandstop amplitude responses for only the positive frequency ranges are shown in fig. 3.1. It is seen that they are specified within the frequency range 0 to $\omega_s/2$. The other parts of the frequency spectrum are periodic repetitions of the baseband which should have no effect on the baseband range, as discussed in section 1.4.

The main design methods for FIR and IIR filters are given in this and the following chapter. Each of the two classes is treated separately and suitable design methods are developed and illustrated.

3.1   FIR filter properties

A unit impulse applied to FIR filter, fig. 2.10, will propagate through the structure resulting in the impulse response given by the coefficients $a_0, a_1, \ldots, a_M$ at discrete

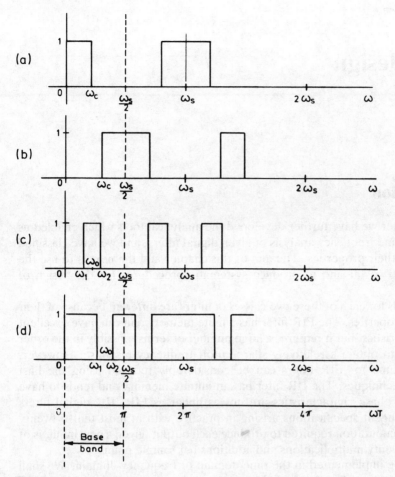

Fig. 3.1   Ideal amplitude responses of (a) lowpass, (b) highpass, (c) bandpass and (d) bandstop filters

times $n = 0, 1, \ldots, M$. Therefore, we may change the notation from a_n to $h(n)$, which is the usual notation for the impulse response, and write the transfer function (2.27) as

$$H(z) = \sum_{n=0}^{M} h(n)z^{-n} \tag{3.1}$$

and the corresponding output/input difference equation is

$$y(k) = \sum_{n=0}^{M} h(n)x(k-n)$$

It has been shown, in sections 1.4 and 2.3, that the frequency response $H(\omega)$ of a discrete-time sequence is periodic and continuous in frequency. Therefore, by setting $z = e^{j\omega T}$ in equation 3.1, and extending the periodic repetition of $H(\omega)$ over all frequencies, positive and negative, we obtain

$$H(\omega) = \sum_{n=-\infty}^{\infty} h(n)e^{-jn\omega T} \tag{3.2}$$

This expression being periodic is suitable for application of Fourier series, so that $h(n)$ is given by

$$h(n) = \frac{1}{\omega_s} \int_{-\omega_s/2}^{\omega_s/2} H(\omega)\, e^{jn\omega T}\, d\omega \qquad (3.3)$$

where the integration is over the baseband range $\omega_s/2 \leqslant \omega \leqslant \omega_s/2$.

Another important property of the FIR filters is obtained from the transfer function (3.1) by setting $z = e^{j\omega T}$ as follows:

$$H(\omega) = \sum_{n=0}^{2M} h(n)\, e^{-jn\omega T} = h(0) + h(1)\, e^{-j\omega T} + h(2)\, e^{-j2\omega T} + \ldots + h(M)\, e^{-jM\omega T} \ldots$$

$$+ h(2M-2)\, e^{-j(2N-2)\omega T} + h(2M-1)\, e^{-j(2M-1)\omega T}$$
$$+ h(2M)\, e^{-j2M\omega T}$$

where in order to simplify we have taken the upper limit as $2M$. Extracting the middle term factor

$$H(\omega) = e^{-jM\omega T}[h(0)\, e^{jM\omega T} + h(1)\, e^{j(M-1)\omega T} + h(2)\, e^{j(M-2)\omega T} + \ldots + h(M)$$
$$+ \ldots h(2M-2)\, e^{-j(M-2)\omega T} + h(2M-1)\, e^{-j(M-1)\omega T} + h(2M)\, e^{-jM\omega T}]$$

The function within the square bracket can become a real function of ωT for two special cases:

(i) $\quad h(0) = h(2M),\ h(1) = h(2M-1),\ h(2) = h(2M-2),$ etc. $\qquad\qquad$ (3.4)
(ii) $\quad h(0) = -h(2M),\ h(1) = -h(2M-1),\ h(2) = -h(2M-2)\ldots,\ h(M) = 0$ $\quad$ (3.5)

In the first case (symmetrical impulse response), we have

$$H(\omega) = e^{-jM\omega T}[h(M) + 2 \sum_{n=0}^{M-1} h(n) \cos(M-n)\omega T] \qquad (3.6)$$

and in the second case (antisymmetrical impulse response), we have

$$H(\omega) = je^{-jM\omega T}[2 \sum_{n=0}^{M-1} h(n) \sin(M-n)\omega T] \qquad (3.7)$$

In both cases the phase response (θ) is a linear function of frequency. In the first case $\theta = -M\omega T$, and in the second case $\theta = \pi/2 - M\omega T$. Furthermore, they are independent of the coefficients $h(n)$. The amplitude response is a cosine and sine function for the first and second case respectively, and it can be adjusted by choice of coefficients $h(n)$. These arrangements have also been used in the past with the transversal filter networks described by Linke(14).

We have denoted the upper limit by $2M$, hence the filter length is $2M+1$. In technical literature generally accepted notation for the upper limit is $M-1$, hence the filter length M (odd or even number). We are dealing here only with an odd number of coefficients, but details for odd and even length filters with symmetrical or asymmetrical responses can be found in Proakis & Manolakis(11) or in Parks & Burrus(15).

3.2 Design procedure

The periodic frequency response $H(\omega)$ and the corresponding impulse sequence $h(n)$ are linked by the Fourier pair of equations 3.2 and 3.3. The advantage of the Fourier series application, in this case, is that for a given $H(\omega)$ we immediately obtain the impulse response $h(n)$ by solving equation 3.3, and therefore the coefficients for the filter $a_n = h(n)$. In this procedure it is usually assumed that $H(\omega)$ has an idealized frequency response, as shown in fig. 3.1, which greatly simplifies solving the integral in equation 3.3. The solution for $h(n)$ contains an infinite number of terms due to the infinitely sharp (discontinuous) cutoff of the ideal filter characteristic. To make the solutions practically realizable $h(n)$ must be truncated to a finite length sequence. This in turn causes a spread of the frequency response into the stopband, and to control this we introduce a suitable window function to modify the $h(n)$ sequence. These points will be illustrated below on an ideal lowpass filter, and also in section 3.3 where we introduce window functions. The transformations from lowpass to other filter types are also discussed in this section.

Example 3.1
Consider the ideal lowpass frequency response characteristic with cutoff frequency ω_c within the range $\pm\omega_s/2$ as shown in fig. 3.2. Applying equation 3.3, we have

$$h(n) = \frac{1}{\omega_s} \int_{-\omega_s/2}^{\omega_s/2} H(\omega)\,e^{jn\omega T}\,d\omega = \frac{1}{\omega_s} \int_{-\omega_c}^{\omega_c} e^{j(n-\lambda)\omega T}\,d\omega$$

with the solution

$$h(n) = \frac{\omega_c T}{\pi}\,\frac{\sin(n-\lambda)\omega_c T}{(n-\lambda)\omega_c T} \tag{3.8}$$

where $n = \lambda, \lambda \pm 1, \lambda \pm 2, \ldots$. This is an infinite sequence of type $(\sin x)/x$ centred at λ, which represents the filter delay.

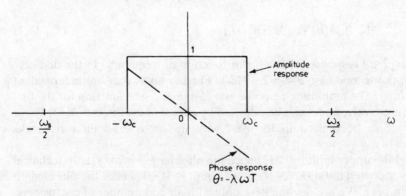

Fig. 3.2 Ideal lowpass filter frequency response

Example 3.2
To simply the analysis, the phase characteristic is often neglected by assuming zero delay, i.e. $\lambda = 0$. We follow the same practice but introduce λ again at a later stage.

For the cutoff $\omega_c = \omega_s/8$, and $\lambda = 0$, equation 3.8 gives us the following impulse response:

$$h(n) = \tfrac{1}{4}\left(\frac{\sin n\pi/4}{n\pi/4}\right) \tag{3.9}$$

where $n = 0, \pm 1, \pm 2, \ldots, \pm\infty$. This is an infinite series of terms which cannot be realized because it would require an infinite number of filter coefficients and an infinite delay. To make it realizable we truncate the series to 21 terms as shown in fig. 3.3(a).

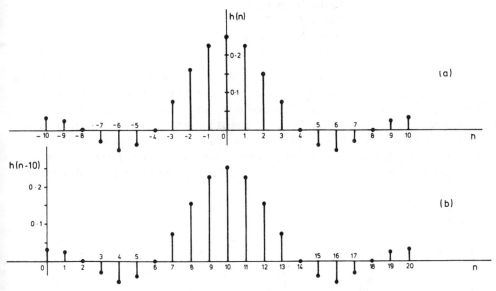

Fig. 3.3 (a) Truncated impulse response for ideal lowpass filter with $\omega_c = \omega_s/8$, (b) as (a) but with finite delay

Introducing now a finite delay of ten units of time (i.e. $\lambda = 10$), we obtain $h(n-10)$ series with values given in the first column of Table 3.1, and also shown in fig. 3.3(b). This is now a realizable (causal) impulse response since it is zero for negative values of n. The values in the second and third columns will be referred to in the next section.

Because the impulse response is symmetrical around $\lambda = 10$, we can calculate its

Table 3.1

$h(n-10)$	R	H	K
$h(0) = h(20)$	0.03183099	0.00254648	0.00036542
$h(1) = h(19)$	0.02500879	0.00256375	0.00113711
$h(2) = h(18)$	0	0	0
$h(3) = h(17)$	−0.03215415	−0.00866936	−0.00637989
$h(4) = h(16)$	−0.05305165	−0.02110671	−0.01698820
$h(5) = h(15)$	−0.04501582	−0.02430854	−0.02092616
$h(6) = h(14)$	0	0	0
$h(7) = h(13)$	0.07502636	0.06079995	0.05758099
$h(8) = h(12)$	0.15915494	0.14517283	0.14168995
$h(9) = h(11)$	0.22507908	0.22001165	0.21867624
$h(10)$	0.25	0.25	0.25

Column R: truncated impulse response series (rectangular window)
Column H = Column R × Hamming window weights
Column K = Column R × Kaiser window weights

frequency amplitude response from equation 3.6. Expanding this equation and using the values from Table 3.1, we obtain the following expression for the amplitude response:

$$
\begin{aligned}
|H(m)|_R = 0.25 &+ 0.45015816 \cos m\pi \\
&+ 0.31830988 \cos 2m\pi \\
&+ 0.15005272 \cos 3m\pi \\
&- 0.09003164 \cos 5m\pi \\
&- 0.10610330 \cos 6m\pi \\
&- 0.06430830 \cos 7m\pi \\
&+ 0.05001758 \cos 9m\pi \\
&+ 0.06366198 \cos 10m\pi
\end{aligned} \tag{3.10}
$$

where we have introduced $\omega T = m\pi$ with $0 \le m \le 1$. In fact $m = \omega/(\omega_s/2)$ is the normalized frequency with respect to half the sampling frequency. The values of coefficients in Table 3.1 have been calculated using an electronic pocket calculator, but the amplitude response $|H(m)|$ has been calculated using the computer program given in section 4 of the appendix. The result is shown in fig. 3.4 by the graph marked as 21 coefficients. To illustrate the effect of the number of coefficients on $|H(m)|$ response, we have performed calculations for 11 and 31 coefficients and these too are marked in fig. 3.4.

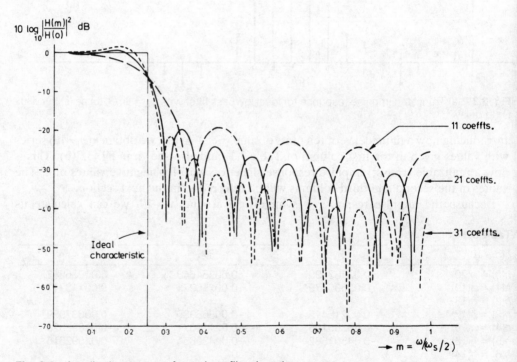

Fig. 3.4 Amplitude response for various filter lengths

Before proceeding to the next section, in which window functions are introduced, we briefly discuss the relationships between impulse responses of different types of filters. It can be shown that the unit-sample or impulse response of the highpass filter with the amplitude response shown in fig. 3.1(b), with $\omega_c = (\omega_s/2) - (\omega_c)_{LP}$, is given by

$$h(n)_{\text{HP}} = (-1)^n h(n)_{\text{LP}} \quad \text{for } n = 0, \pm1, \pm2, \dots \tag{3.11}$$

where $h(n)_{\text{LP}}$ is the lowpass unit-sample response.

For the bandpass filter of fig. 3.1(c), we have

$$h(n)_{\text{BP}} = (2 \cos n\omega_0 T) h(n)_{\text{LP}} \quad \text{for } n = 0, \pm1, \pm2, \dots \tag{3.12}$$

where ω_0 is the bandpass centre frequency, and $\omega_1 = \omega_0 - \omega_c$, $\omega_2 = \omega_0 + \omega_c$, where ω_c refers to the lowpass filter as in fig. 3.1(a). The unit-sample response of the bandstop filter in fig. 3.1(d) is related to the bandpass filter in the following way:

$$h(0)_{\text{BS}} = 1 - h(0)_{\text{BP}}$$
$$h(n)_{\text{BS}} = -h(n)_{\text{BP}} \quad \text{for } n = \pm1, \pm2, \dots \tag{3.13}$$

Applying the above relationships to the lowpass filter unit-sample response of fig. 3.3, we obtain the unit-sample responses for the highpass, bandpass and bandstop filters shown in fig. 3.5.

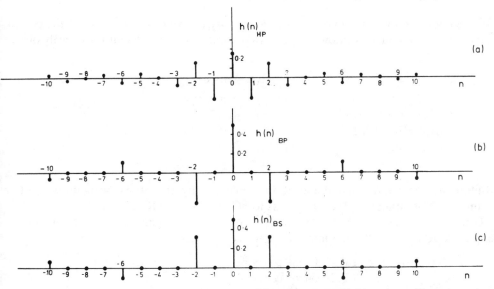

Fig. 3.5   Unit-sample response for (a) highpass, (b) bandpass and (c) bandstop filters based on the lowpass of Fig. 3.3

3.3   Design modification using windows

The truncation of an infinite time series, discussed in the previous section, corresponds to the application of a rectangular window function defined as

$$w_{\text{R}}(n) = \begin{cases} 1 & |n| \le N \\ 0 & |n| > N \end{cases}$$

The frequency response of a truncated time series can be improved by using a number of window functions which modify the unit-sample response $h(n)$ in a prescribed way by multiplication, i.e. $h(n) \times w(n)$. We discuss briefly two typical cases known as the generalized Hamming and Kaiser window functions. The generalized Hamming window function is given by

$$w_H(n) = \begin{cases} \alpha + (1-\alpha)\cos(n\pi/N) & |n| < N \\ 0 & |n| > N \end{cases} \tag{3.14}$$

where $0 \le \alpha \le 1$. If $\alpha = 0.54$ the window is called a Hamming window, and if $\alpha = 0.50$ it is called a Hanning window. An application of the Hamming window is illustrated in the example shown below.

Example 3.3
This is an extension of example 3.2 where $M = 21$, hence $N = (M-1)/2 = 10$. Choosing $\alpha = 0.54$, and for $N = 10$, equation 3.14 becomes

$$w_H(n) = 0.54 + 0.46\cos(n\pi/10)$$

This function is symmetrical around the value of $w_H(0)$ which corresponds to the central point of the rectangular window in fig. 3.3(a). Using a pocket calculator we easily obtain the following values:

$w_H(0) = 1$ $w_H(6) = 0.39785218$
$w_H(1) = 0.97748600$ $w_H(7) = 0.26961878$
$w_H(2) = 0.91214782$ $w_H(8) = 0.172$
$w_H(3) = 0.81038122$ $w_H(9) = 0.10251400$
$w_H(4) = 0.678$ $w_H(10) = 0.08$
$w_H(5) = 0.54$

Multiplying the values of column R in Table 3.1 by the above weights we obtain column H. Note than $w_H(0)$ corresponds to $h(10)$ in column R of Table 3.1.

Using the values of column H, in equation 3.6, we obtain the expression for the amplitude response with Hamming window, as

$$\begin{aligned}
|H(m)|_H = \ &0.25 + 0.44002330\cos m\pi \\
&+ 0.29034566\cos 2m\pi \\
&+ 0.12159990\cos 3m\pi \\
&- 0.04861708\cos 5m\pi \\
&- 0.04221342\cos 6m\pi \\
&- 0.01733872\cos 7m\pi \\
&+ 0.00512750\cos 9m\pi \\
&+ 0.00509296\cos 10m\pi
\end{aligned} \tag{3.15}$$

with the same comments as given for equation 3.10. The results of computer calculations as in section 4 of the appendix for the ratio

$$20\log_{10}|H(m)/H(0)|)\,(\text{dB})$$

where $H(m)$ is given by equation 3.15, are shown in fig. 3.6 under the Hamming window, where the result for the rectangular window of example 3.2 is also shown for comparison.

The second window we discuss is a family of weighting functions proposed by Kaiser(16),

$$w_K(n) = \begin{cases} \dfrac{I_0[\beta\sqrt{(1-(n/N)^2)}]}{I_0(\beta)} & |n| \le N \\ \\ 0 & |n| > N \end{cases} \tag{3.16}$$

where I_0 is the modified Bessel function of the first kind and zero order. The parameter β specifies its frequency response in terms of the main lobe width and the side lobe level. Large values of β correspond to a wider main lobe width and smaller side lobe levels. The normal range of β is $4 \le \beta \le 9$, corresponding to a range of side lobe peak heights of 3.1% down to 0.047%.

The weighting function given in equation 3.16 is symmetrical around $w_H(0)$, so it is sufficient to calculate its value for $n = 0, 1, \ldots, N$. These values are then used to modify the $h(n)$ response.

Example 3.4
This is again an extension of example 3.2, but now using the Kaiser window. Choosing the value $\beta = 2\pi$ in equation 3.16, and $N = 10$ as in example 3.2, we have

$$w_K(n) = \frac{I_0[2\pi\sqrt{(1-(n/10)^2)}]}{I_0(2\pi)}$$

where $n = 0, 1, 2, \ldots, 10$. The difficulty is that the tabulated Bessel functions in mathematical handbooks do not give $I_0(x)$ for all values of x, so employing the power series expansion

$$I_0(x) = 1 + \sum_{m=1}^{M} \left(\frac{(x/2)^m}{m!} \right)^2 \tag{3.17}$$

and evaluating this expression to within specified accuracy using the computer program proposed by Kaiser and given in section 4 of the appendix (also in Rabiner & Gold(13), p. 103), we obtain the values for the weights given in Table 3.2. The last column of Table 3.2 gives the upper limit M used in the sum of equation 3.17, based on the criterion that the last term in the series is less than 1×10^{-8}.

Multiplying the values of column R in Table 3.1 by the weights given in Table 3.2 we obtain the column K. We note again that $w_K(0)$ corresponds to $h(10)$ in column R of

Table 3.2 Weights for nonrecursive filter using Kaiser window function

n	$w_k(n)$	M
0	1.0000000000	15
1	0.9715529534	14
2	0.8902642058	14
3	0.7674768355	14
4	0.6195788531	14
5	0.4648623206	13
6	0.3202200325	13
7	0.1984156504	12
8	0.1064793789	11
9	0.0454683321	9
10	0.0114799346	1

Table 3.1. Using the values of column K in equation 3.6 we obtain the following expression for the amplitude response.

$$
\begin{aligned}
|H(m)|_K = 0.25 &+ 0.43735248 \cos m\pi \\
&+ 0.28337990 \cos 2m\pi \\
&+ 0.11516198 \cos 3m\pi \\
&- 0.04185232 \cos 5m\pi \\
&- 0.03397640 \cos 6m\pi \\
&- 0.01275978 \cos 7m\pi \\
&+ 0.00227422 \cos 9m\pi \\
&+ 0.00073084 \cos 10m\pi
\end{aligned}
\tag{3.18}
$$

with explanations as given for equation 3.10. The response has been calculated as

$$20\log_{10}|H(m)/H(0)|\,(\mathrm{dB})$$

using the computer program of section 4 of the appendix, and shown in fig. 3.6 under the Kaiser window.

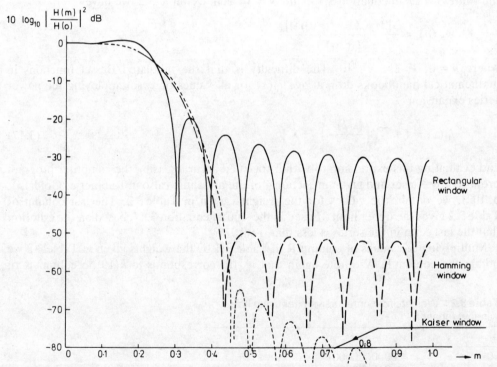

Fig. 3.6 Amplitude response using rectangular window, Hamming window and Kaiser window functions

Comparing fig. 3.4 with fig. 3.6 we see that the window shaping of filter coefficients is more powerful than the extension of the number of coefficients from 21 to 31 in reducing the stopband response. However, although the window functions reduce the stopband sidelobes, this is done at the expense of an increase in the width of the filter transition band.

The multiplication of the impulse response by the window function in the time-domain, $h(n) \times w(n)$, corresponds to the convolution in the frequency-domain, i.e. $H(\omega) * W(\omega)$. The window weighting function $w(n)$ is chosen so that $W(\omega)$ is narrow-band. The major effect of the window is that discontinuities in $H(\omega)$ become transition bands on either side of discontinuity. Frequency responses of various window functions can be found, for example, in Oppenheim & Schaffer(10), section 5.5, and Rabiner & Gold(13), section 3.12.

3.4   Design by frequency sampling method

In the previous method equation 3.3 has been used to obtain the filter coefficients $h(n)$. The filter function, $H(\omega)$ in equation 3.3, is usually assumed to have the ideal shape as in fig. 3.1, hence easy for integration. If $H(\omega)$ is not a simple function, or if it is obtained experimentally, solutions of equation 3.3 would not be easy.

The alternative is to take samples of $H(\omega)$. For this purpose it is convenient to modify equation 3.3 as follows: set $T = 1$, i.e. $f_s = 1$ Hz, hence $\omega_s = 2\pi/T$ becomes 2π. Assume the FIR filter length M (before denoted as $2M + 1$), and use it to define the frequency sampling interval as $2\pi/M$. Now the sampled frequency is given by $\omega_k = (2\pi/M)k$, where k is an integer within the range $-(M-1)/2 \leq k \leq (M-1)/2$, as shown in fig. 3.7. Using these quantities in equation 3.3, and changing the integration to summation, we obtain

$$h(n) = \frac{1}{M} \sum_{k=-(M-1)/2}^{(M-1)/2} H(k) e^{j(2\pi/M)kn} \tag{3.19}$$

Since the magnitude response of the desired frequency response is an even (symmetrical) function of k, i.e. $H(k) = H(-k)$, we can express equation 3.19 as

$$h(n) = \frac{1}{M} \left[H(0) + 2 \sum_{k=1}^{(M-1)/2} H(k) \cos\left(\frac{2\pi}{M}\right) kn \right] \tag{3.20}$$

which is more convenient to use in designs.

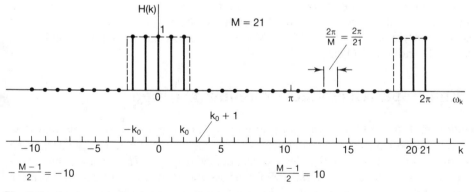

Fig. 3.7   An example for the frequency sampling method

Example 3.5
We consider the same case as specified in example 3.2, for FIR of length $M = 21$. Since the cutoff is $\omega_s/8 = 0.25\pi$, we have only five nonzero samples of $H(k)$, in the interval $-10 \leq k \leq 10$, as shown in fig. 3.7. Equation 3.20 for this case gives

$$h(n) = \frac{1}{21}\left[H(0) + 2 \sum_{k=1}^{2} H(k) \cos(2\pi/21)kn \right]$$

where $n = 0, \pm1, \pm2, \ldots, \pm10$. Results are:

$h(0)$ = 0.238095 (0.25);	$h(\pm1)$ = 0.217315 (0.225079)
$h(\pm2)$ = 0.161103 (0.159155);	$h(\pm3)$ = 0.085806 (0.075026)
$h(\pm4)$ = 0.0126 (0);	$h(\pm5)$ = -0.039438 (-0.045016)
$h(\pm6)$ = 0.059379 (-0.053052);	$h(\pm7)$ = -0.047619 (-0.032154)
$h(\pm8)$ = 0.015078 (0);	$h(\pm9)$ = 0.021192 (0.025008)
$h(\pm10)$ = 0.04445 (0.031831)	

The values in brackets are from Table 3.1 for column R. Differences between $h(n)$ for the previous (windowing) method and this (frequency sampling) method vary from 0.002 up to 0.016. In windowing method we have decreased the ripple at the price of widening the transition band, fig. 3.6. Similarly, in this case of the frequency sampling method, we can decrease the ripple by widening the transition band. This means instead of specifying $H(k_0) = 1$ and $H(k_0+1) = 0$, as in fig. 3.7, the transition band is increased by setting $H(k_0) = 1$, $0 \le H(k+1) \le 1$, and $H(k_0+2) = 0$. The value of $H(k_0+1)$ can be optimized, in some sense, to reduce the ripple, Oppenheim & Schaffer(10). Other types of impulse response and frequency sampling intervals are discussed in Proakis and Manolakis(11).

Equation (3.1) for M coefficients is

$$H(z) = \sum_{n=0}^{M-1} h(n)z^{-n} \tag{3.21}$$

Substituting equation 3.19 for $h(n)$, one can obtain the transfer function directly in terms of sampled frequency response as follows

$$H(z) = \left(\frac{1-z^{-M}}{M}\right) \sum_{k=0}^{M-1} \frac{H(k)}{1 - e^{j(2\pi/M)k}z^{-1}} \tag{3.22}$$

The above indicates a special kind of implementation consisting of an all-zero, or a comb filter, cascaded by a parallel bank of single-pole filters with resonant frequencies at $p_k = \exp[j(2\pi/M)k]$, $k = 0, 1, \ldots, M-1$. More information can be found in Proakis and Manolakis(11) and in Kuc(17).

3.5   Equiripple (Chebyshev or minimax) design

The window method and the frequency sampling method are relatively simple, but they have no control over the critical frequencies ω_p, ω_s, fig. 3.8. In the equiripple design the approximation error between the desired frequency response $H_d(\omega)$, and the actual frequency response, $H(\omega)$, is spread evenly across the passband and evenly over the stopband, and the maximum error is minimized (hence the term minimax design).

The main design points of this method are as follows.
From equation (3.21), for $z = e^{j\omega}$, the actual frequency response can be written as

$$H(\omega) = \sum_{m=0}^{L=(M-1)/2} \alpha(m) \cos m\omega \tag{3.23}$$

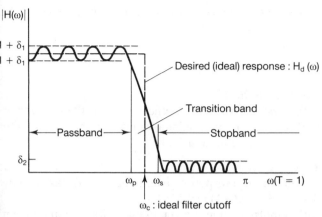

ω_c : ideal filter cutoff

δ_1, δ_2 : passband and stopband ripples.

Fig. 3.8 Lowpass filter specification

where $m = L - n$, $\alpha(0) = h(L)$, $\alpha(m = 1, 2, \ldots, L) = 2h(L - m)$.
The error is given by

$$E(\omega) = H_d(\omega) - H(\omega)$$

$$= H_d(\omega) - \sum_{m=0}^{L} \alpha(m) \cos m\omega \tag{3.24}$$

and it is to be minimized over parameters α. The best (Chebyshev) approximation to $H_d(\omega)$ requires that the error function $E(\omega)$ has at least $L + 2$ extremal frequencies in the interval $[0, \pi]$. This results in a set of equations for $(L + 1)$ α-parameters, and one for $\delta = \max |E(\omega)|$. The problem is that we don't know the set of extremal frequencies but the solution is given in terms of the Remez exchange algorithm. This is quite a complex design procedure requiring the use of computers. The theoretical details can be found in Proakis and Manolakis(11) and the computer program for this method is given in Parks and Burrus(15).

3.6 Problems

3.1 The analogue differentiator is described by the transfer function

$$H(s) = sT$$

where s is the Laplace variable. Design a nonrecursive digital differentiator to simulate the above function up to $f_s/2$, where $f_s = 1/T$ is the sampling frequency. Calculate the first 19 coefficients of the nonrecursive filter.
Note: For solving this problem, it is useful to know $\int x e^{jx} dx = e^{jx}(1 - jx)$.

3.2 (a) Calculate the magnitude of the frequency response for three and five middle terms in the solution of problem 3.1.
(b) Compare the above two cases graphically with the ideal differentiator $|H(\omega)| = \omega T$ over the range of frequencies 0 to $\omega_s/2$. Observe the improvement made by taking five instead of three terms.

3.3   (a) Derive the expression for the coefficients of a nonrecursive filter which is to have frequency response

$$H(\omega) = \begin{cases} 1 & 0 \le \omega \le \omega_s/4 \\ 0 & \text{elsewhere} \end{cases}$$

where ω_s is the sampling frequency in rad s^{-1}.

(b) Calculate the values of the first twelve filter coefficients from the expression derived in problem 3.3(a). Determine their modified values for the case of the generalized Hamming window function, equation 3.14.

3.4   Calculate the magnitude of the frequency response:

(a) for the lowpass filter with the impulse response given in fig. 3.3(a), taking only the three terms $h(0) = 0.25$ and $h(1) = h(-1) = 0.23$;

(b) for the highpass filter with impulse response given in fig. 3.5(a) obtained from the lowpass case using equation 3.11, taking only the three terms $h(0) = 0.25$ and $h(1) = h(-1) = -0.23$. Plot and compare responses of (a) and (b) over the frequency range $-\omega_s/2$ to $\omega_s/2$.

3.5   Calculate the magnitude of the frequency response:

(a) for the bandpass filter with the impulse response given in fig. 3.5(b), obtained from the lowpass case using equation 3.12, taking only the first three nonzero terms $h(0) = 0.50$ and $h(-2) = h(2) = -0.32$;

(b) for the bandstop filter with the impulse response given in fig. 3.5(c), obtained from the above bandpass case using equation 3.13, taking only the first three nonzero terms $h(0) = 0.50$ and $h(-2) = h(2) = 0.32$.

Plot and compare responses of (a) and (b) over the frequency range $-\omega_s/2$ to $\omega_s/2$. Compare these graphs with the ones of problem 3.4.

3.6   A lowpass filter, fig. 3.2, is specified with the cutoff frequency $\omega_c T = \pi/2$ ($T = 1$). Ignoring the phase in fig. 3.2 design a FIR filter of length $M = 13$, and determine its weights by (i) the window method, (ii) the frequency-sampling method.

4

IIR filter design

4.0 Introduction

The division of digital filters into FIR and IIR types has been discussed earlier, particularly in section 2.6. The IIR filter structure and the form of its transfer function dictate design approaches different from those used in FIR filter design. Rewriting equation 2.4 gives the general recursive filter transfer function as

$$H(z) = \frac{\sum_{i=0}^{M} a_i z^{-i}}{1 + \sum_{i=1}^{N} b_i z^{-i}} \tag{4.1}$$

The aim in IIR filter design is to determine the filter coefficients, a_i and b_i, such that the filter specifications are satisfied. One could use Fourier series, as for FIR filters, but (a_i, b_i) are not related to $h(n)$ in a simple way.

There are two basic approaches to this design problem. The first, a direct approach, is to determine the coefficients of the digital filter by some computational procedure directly from the filter specifications, and is described by Bogner & Constantinides(18), chapter 5. The second basic approach, to be considered here, is to determine the coefficients in an indirect way from the analogue (i.e. continuous-time) filters. This method consists of two parts:

(i)   determination of a suitable analogue filter transfer function $H(s)$ which meets the required filter specification;
(ii) digitalization of this analogue filter.

The direct method is implicit in the digital simulation of analogue filters: in this case the analogue filter is already known and only digitalization is required. Regarding design approach based on analogue filters, it is perhaps worthwhile to point out also that the recursive filter is a natural counterpart of the analogue filter as seen in section 1.1.

Digital filter characterization is specified in the frequency range $0 \leq \omega \leq \omega_s/2$, where ω_s is the sampling frequency, as discussed in section 3.0 and illustrated in fig. 3.1. We consider two main methods of IIR filter design, known as the impulse invariant method and the bilinear z-transform, presented in sections 4.1 and 4.2 respectively. These design techniques, developed for the lowpass filter, are extended to other types of filters by means of the frequency transformations discussed in section 4.3. Some general comments on accuracy concerning both the FIR and IIR filters are given in section 4.4. Various examples are given throughout sections of this chapter, and additional information and exercises are given in section 4.5.

47

4.1   The impulse invariance method

As pointed out in the introduction, we start with the analogue filter transfer function $H(s)$ whose general form is given by

$$H(s) = \frac{\sum_{n=0}^{M} c_n s^n}{\sum_{n=0}^{N} d_n s^n} \tag{4.2}$$

If $N > M$, i.e. if the degree of the polynomial in the denominator is larger than the degree of the numerator, and if the poles of $H(s)$ are simple, then its partial-fractions form can be written as

$$H(s) = \sum_{n=1}^{N} \frac{A_n}{s + s_n} \tag{4.3}$$

where A_n are real or complex constants.

It is sufficient for our analysis to consider a typical simple pole term at $s = s_1$, for which we have

$$H_1(s) = \frac{A_1}{s + s_1} \tag{4.4}$$

The corresponding impulse response is

$$h_1(t) = A_1 e^{-s_1 t} \tag{4.5}$$

and its sampled form is written directly as

$$h_1(kT) = A_1 e^{-s_1 kT} \tag{4.6}$$

where T is the sampling interval and k is an integer. The sampled impulse frequency response is given by

$$H_{1s}(s) = T \sum_{k=0}^{\infty} h_1(kT) e^{-skT} \tag{4.7}$$

where s is the complex frequency, and substituting equation 4.6 for $h_1(kT)$, we obtain an infinite geometric series whose sum produces the result

$$H_{1s}(s) = \frac{A_1 T}{1 - e^{-s_1 T} e^{-sT}} \tag{4.8}$$

Applying now the standard z-transformation $z = e^{sT}$ from equation 1.19, we obtain the discrete-time (or sampled-data) transfer function

$$H_1(z) = \frac{A_1 T}{1 - e^{-s_1 T} z^{-1}} \tag{4.9}$$

The impulse invariant method is also known as the standard z-transform method because of the application of $z = e^{sT}$. We already know, from section 1.4 and section 3 of the appendix, that the dixcrete-time transfer function 4.8 or 4.9 is a periodic function with period $\omega_s = 2\pi/T$.

The result equation 4.9 as a single term of equation 4.3 can be extended to n terms as follows:

$$H(s) = \sum_{i=1}^{N} \frac{A_i}{s+s_i} \rightarrow \sum_{i=1}^{N} \frac{A_i T}{1-e^{-s_i T}z^{-1}} = H(z) \tag{4.10}$$

The simple one-pole analogue filter is transformed into a digital filter in the following way:

$$\frac{a}{s+a} \rightarrow \frac{aT}{1-e^{-aT}z^{-1}} \tag{4.11}$$

and for second-order system

$$\frac{s+a}{(s+a)^2+b^2} \rightarrow T\frac{1-e^{-aT}(\cos bT)z^{-1}}{1-2e^{-aT}(\cos bT)z^{-1}+e^{-2aT}z^{-2}} \tag{4.12}$$

$$\frac{b}{(s+a)^2+b^2} \rightarrow T\frac{e^{-aT}(\sin bT)z^{-1}}{1-2e^{-aT}(\cos bT)z^{-1}+e^{-2aT}z^{-2}} \tag{4.13}$$

The last two relationships are obtained by expanding the left-hand-side into partial-fractions and then applying equation 4.11 to each individual term. Combining the two terms we obtain the right-hand-side of equations 4.12 and 4.13.

Example 4.1
We want to design a discrete-time filter of IIR type with the frequency response of an analogue single-action RC filter whose transfer function is given by

$$H(s) = \frac{\omega_c}{s+\omega_c} \tag{4.14}$$

where $\omega_c = 1/RC$ defines the -3dB cutoff point.
   Applying the transformation of equation 4.11 to equation 4.14, we obtain the digital transfer function

$$H(z) = \frac{\omega_c T}{1-z^{-1}e^{-\omega_c T}} \tag{4.15}$$

The discrete-time filter structure is obtained from the above equation by expressing $H = Y/X$, where Y and X refer to the filter output and input respectively. It is easy to show that the filter structure is given by the difference equation

$$y(k) = (\omega_c T)x(k) + e^{-\omega_c T}y(k-1) \tag{4.16}$$

which is shown in fig. 4.1 (see also fig. 1.2(a)).

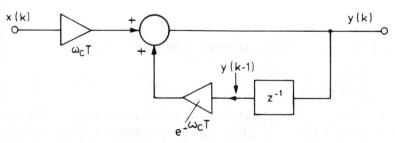

Fig. 4.1   Recursive digital structure equivalent to analogue single-section RC filter

To compare the frequency responses of the above filters we have chosen $\omega_c = 2\pi \times 10^3$ rad s^{-1}, $\omega_s = 2\pi \times 10^4$ rad s^{-1}, for which $\omega_c T = 0.628$. With these values, the analogue filter magnitude squared has been arranged in the form

$$|H(n)|^2 = \frac{1}{1 + n^2/16} \tag{4.17}$$

obtained by the choice of the frequency steps $\Delta f = 250$ Hz. The function $10\log_{10}|H(n)/H(0)|^2$ has been calculated, using a computer, for $n = 0, 1, 2, 3, \ldots, 40$ which covers the frequency range of $40\Delta f = 10$ kHz. The result is shown in fig. 4.2, graph (a).

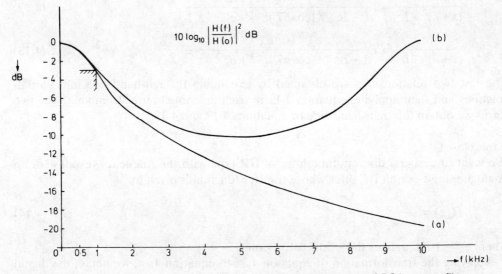

Fig. 4.2   Frequency response for (a) analogue and (b) digital version of RC lowpass filter

With $\omega_c T = 0.628$, the digital filter magnitude squared response for equation 4.15 can be arranged as

$$|H(m)|^2 = \frac{0.394}{1.285 - 1.068 \cos m\pi} \tag{4.18}$$

where we have introduced $\omega T = m\pi$. The function $10\log_{10}|H(m/H(0)|^2$ has been calculated from $m = 0$ to $m = 2$ in steps of 0.05, section 5 of the appendix. The result obtained is shown in fig. 4.2, graph (b), for a comparison with the analogue filter graph (a).

Example 4.2

We determine a digital filter for the three-pole Butterworth lowpass filter

$$H(s) = \frac{s_1 s_2 s_3}{(s + s_1)(s + s_2)(s + s_3)} \tag{4.19}$$

where $s_1 = \omega_c$, $s_2 = \frac{1}{2}(1 + j\sqrt{3})\omega_c$, $s_3 = \frac{1}{2}(1 - j\sqrt{3})\omega_c$; ω_c is the cutoff frequency defined by $|H(\omega_c)| = 0.707$, references (10), (11), (17). Expanding equation 4.19 into partial-fractions we have

$$H(s) = \frac{A_1}{s+s_1} + \frac{A_2}{s+s_2} + \frac{A_3}{s+s_3} \tag{4.20}$$

where $A_1 = \omega_c$, $A_2 = -\frac{1}{2}(1-j.1/\sqrt{3})\omega_c$, and $A_3 = -\frac{1}{2}(1+j.1/\sqrt{3})\omega_c$.

Applying the transformation given by equation 4.11 to each term in equation 4.20 we obtain the following result:

$$H(z) = \frac{\omega_c T}{1-e^{-\omega_c T}z^{-1}} + \frac{[-1+f(\omega_c T)z^{-1}]\omega_c T}{1-[2e^{-\omega_c T/2}\cos(\frac{\sqrt{3}}{2}\omega_c T)]z^{-1}+e^{-\omega_c T}z^{-2}} \tag{4.21}$$

where $f(\omega_c T) = e^{-\omega_c T/2}[\cos(\frac{\sqrt{3}}{2}\omega_c T) + \sqrt{\frac{1}{3}}\sin(\frac{\sqrt{3}}{2}\omega_c T)]$. This result is in the form suitable for the parallel realization of fig. 2.9.

To derive the difference equation, and hence the filter structure, we express $H(z)$ as

$$H(z) = \frac{Y(z)}{X(z)} = \frac{Y_1(z)}{X(z)} + \frac{Y_2(z)}{X(z)} \tag{4.22}$$

Comparing equations 4.21 and 4.22, we have

$$H_1(z) = \frac{Y_1(z)}{X(z)} = \frac{\omega_c T}{1-e^{-\omega_c T}z^{-1}}$$

from which we obtain the difference equation

$$y_1(k) = (\omega_c T)x(k) + (e^{-\omega_c T})y_1(k-1) \tag{4.23}$$

Similarly for the second term,

$$y_2(k) = (-\omega_c T)x(k) + [\omega_c Tf(\omega_c T)]x(k-1) + [2e^{-\omega_c T/2}\cos(\omega_c T)]y_2(k-1) \\ - (e^{-\omega_c T})y_2(k-2) \tag{4.24}$$

The complete solution is the sum of equations 4.23 and 4.24

$$y(k) = y_1(k) + y_2(k)$$

defining the digital filter structure shown in fig. 4.3.

We have also calculated magnitude responses for the analogue and digital filter transfer functions given by equations 4.19 and 4.21 respectively. The values $\omega_c = 2\pi \times 10^3 \text{ rad s}^{-1}$, $\omega_s = 2\pi \times 10^4 \text{ rad s}^{-1}$ and $\omega_c T = 0.628$ have been chosen as in example 4.1.

For the analogue filter of equation 4.19, we have

$$|H(n)|^2 = \frac{1}{1+n^6/4096} \tag{4.25}$$

for frequency steps of 250 Hz and $n = 0, 1, 2, 3, \ldots, 40$.

Similarly, for the digital filter of equation 4.21, we have

$$|H(m)|^2 = \frac{0.009 + 0.008\cos m\pi}{5.712 - 8.546\cos m\pi + 3.421\cos 2m\pi - 0.570\cos 3m\pi} \tag{4.26}$$

for $\omega T = m\pi$ and $0 \le m \le 2$, in steps of 0.05.

Using the above equations and the computer program given in section 5 of the appendix, we have calculated the quantity $10\log_{10}|H/H(0)|^2$ for each case. Results are

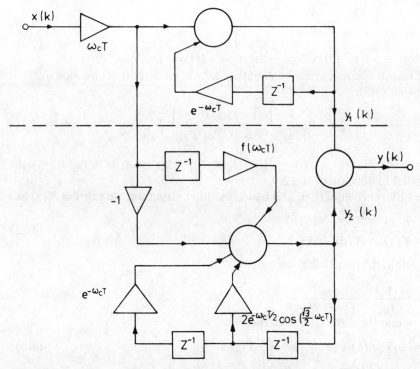

Fig. 4.3   Digital structure for three-pole Butterworth lowpass filter

shown in fig. 4.4(a) for the analogue filter (equation 4.25), and in fig. 4.4(b) for the digital version of this filter (equation 4.26).

There are problems inherent in the design of digital filters by the impulse invariant method. The main one is caused by the interspectra interference (spectrum 'folding') introduced by terms $H(\omega + n\omega_s)$ for $n \neq 0$. If the analogue filter transfer function $H(s)$ is bandlimited to the baseband, i.e. $H(\omega) = 0$ for $\omega > \omega_s/2$, then there is no folding error and the frequency response of the digital filter is identical to that of the original analogue filter. However, when $H(s)$ is a rational function of s, it is not bandlimited and therefore $H(s) \neq H_s(s)$ in the baseband. The magnitude of errors resulting from the folding is directly related to the high frequency asymptotic behaviour of $H(s)$, as seen by comparing graphs (a) and (b) in figs 4.2 and 4.4. To reduce the possibility of such errors, the function $H(s)$ can be modified by adding in cascade a bandlimiting filter to $\omega_s/2$ as mentioned in section 1.4.

The bilinear transformation, next section, overcomes this limitation.

4.2   The bilinear z-transformation method

In the previous section we have dealt with one term in the partial-fraction expansion of an analogue filter given by equation 4.4, which is rewritten here as

$$H(s_a) = \frac{A_1}{s_a + s_1} \tag{4.27}$$

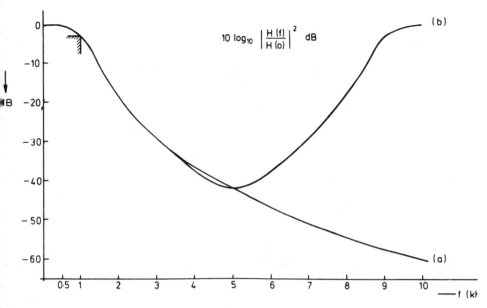

Fig. 4.4 Frequency response for (a) analogue and (b) digital versions of three-pole Butterworth lowpass filter

The variable s_a is used in place of s as a more definite notation for the analogue filter. We introduce now the transformation

$$s_a = \frac{2}{T} \tanh \frac{s_d T}{2} \tag{4.28}$$

where s_d is a new variable whose properties and effects are discussed below. Substituting equation 4.28 into 4.27, and setting $s_a = j\omega_a$ and $s_d = j\omega_d$ we obtain the frequency response

$$H(\omega_d) = \frac{A_1}{s_1 + j\frac{2}{T} \tan \frac{\omega_d T}{2}} \tag{4.29}$$

and the transformation, equation 4.28, becomes

$$\omega_a = \frac{2}{T} \tan \frac{\omega_d T}{2} \tag{4.30}$$

From the above expressions we can make the following observations. First we see that the analogue filter frequency response $H(\omega_a)$ in equation 4.27 tends to zero only as $\omega_a \to \infty$. However, the modified function $H(\omega_d)$ in equation 4.29 goes to zero for finite values of ω_d given by

$$\omega_d T/2 = (2\lambda + 1)\pi/2 \tag{4.31}$$

where $\lambda = 0, \pm 1, \pm 2, \ldots$. Therefore, we see that the transformation given in equation 4.28 compresses the infinite frequency range ω_a into a finite range, and makes the

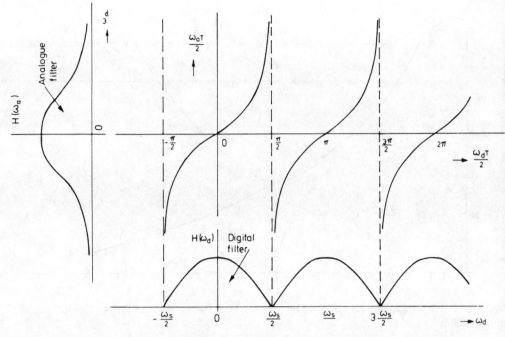

Fig. 4.5   Illustration of spectra compression due to the bilinear z-transformation

modified transfer function given by equation 4.29 periodic. These properties are illustrated in fig. 4.5, where the quantity T has been interpreted, as before, as the reciprocal of the sampling frequency ($T = 1/f_s$). This means that the spectra folding problem is eliminated since the base-band is confined to $\omega_s/2$, but the disadvantage to this is a distorted frequency scale as shown in fig. 4.6. As will be seen in later examples, this frequency scale distortion (or warping) is taken into account in the course of the digital filter design.

Returning to the transformation of equation 4.28, we can write it as

$$s_a = \frac{2}{T} \frac{1 - e^{-s_d T}}{1 + e^{-s_d T}}$$

and applying the standard z-transformation, in this case $z = e^{s_d T}$, we obtain

$$s_a = \frac{2}{T} \frac{1 - z^{-1}}{1 + z^{-1}} \tag{4.32}$$

This relationship, known as the bilinear z-transformation, links the analogue filter variable s_a with the digital filter variable z. To obtain the digital filter transfer function $H(z)$ from a given analogue filter transfer function $H(s)$, we simply substitute from equation 4.32, giving

$$H(z) = H(s) \bigg|_{s = \frac{2}{T} \frac{1 - z^{-1}}{1 + z^{-1}}} \tag{4.33}$$

This method of digital filter design is illustrated in the following examples.

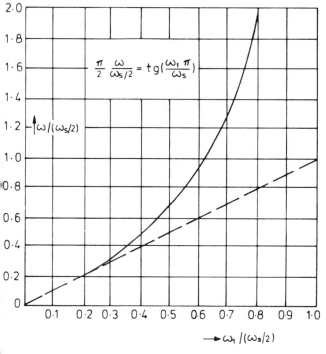

Fig. 4.6 Distorted frequency scale due to the bilinear z-transformation

Example 4.3

Consider the first-order analogue (RC) filter used in example 4.1, whose transfer function is

$$H(s_a) = \frac{\omega_{ac}}{\omega_{ac} + s_a} \tag{4.34}$$

where we added the subscript a to denote the analogue filter. We want to design a digital first-order filter with cutoff frequency ω_{dc}. Then the equivalent analogue filter cutoff frequency is given by equation 4.30, i.e.

$$\omega_{ac} = \frac{2}{T} \tan \frac{\omega_{dc} T}{2} \tag{4.35}$$

The digital filter, as given by equation 4.33, is obtained from equation 4.34 with the application of equation 4.32

$$H(z) = \frac{(\omega_{ac} T/2)(1 + z^{-1})}{(\omega_{ac} T/2 + 1) + (\omega_{ac} T/2 - 1) z^{-1}} \tag{4.36}$$

From the above we obtain, as before, the difference equation

$$y(k) = \frac{\omega_{ac} T/2}{1 + \omega_{ac} T/2} [x(k) + x(k-1)] + \frac{1 - \omega_{ac} T/2}{1 + \omega_{ac} T/2} y(k-1) \tag{4.37}$$

which defines the hardware filter structure.

We calculate the frequency response of this filter by choosing $\omega_{dc} = 2\pi \times 10^4\,\text{rad s}^{-1}$ and $\omega_s = 2\pi \times 10^3\,\text{rad s}^{-1}$, so that $\omega_{dc}\,T = 0.2\pi$ as in example 4.1.

The equivalent analogue filter cutoff is then obtained from equation 4.35, or

$$\omega_{ac}\,T/2 = \tan(\omega_{dc}\,T/2) = \tan(0.1\pi) = 0.325$$

Using this value in equation 4.36, we can calculate the magnitude squared of the frequency response and express it in the following way:

$$|H(m)|^2 = 0.212\frac{1 + \cos m\pi}{2.212 - 1.789\cos m\pi} \tag{4.38}$$

The quantity $10\log_{10}|H/H(0)|$ has been calculated using the computer program in section 5 of the appendix, for $0 \le m \le 2$ in steps of 0.05. The result obtained is plotted in

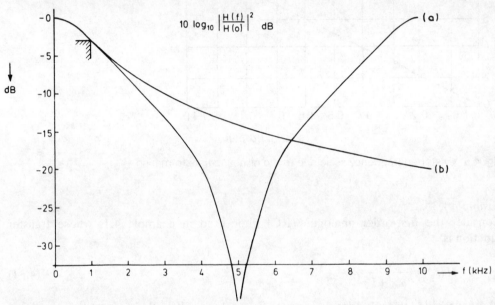

Fig. 4.7 Frequency response for first-order RC filter: (a) digital version with the bilinear z-transformation and (b) original analogue version

fig. 4.7 as graph (a), together with the analogue filter of the same cutoff, graph (b), calculated earlier in example 4.1.

Example 4.4

We design a digital filter, for 10 kHz sampling rate, which is flat to 3 dB in the passband of 0 to 1000 Hz, and which is more than 10 dB down at frequencies beyond 2000 Hz. The filter must be monotonic in passband and stopband. It is known from the analogue filter theory that a Butterworth filter can meet such a specification.

The specified characteristic frequencies are $\omega_{d1}\,T = 2\pi \times 10^3 \times 10^{+4} = 0.2\pi$, and $\omega_{d2}\,T = 0.4\pi$. The corresponding set of analogue frequencies are obtained by means of equation 4.30 as

$$\omega_{a1} = \tan(\omega_{d1}\,T/2) = \tan 0.1\pi = 0.325$$

$$\omega_{a2} = \tan(\omega_{d2}\,T/2) = \tan 0.2\pi = 0.726$$

The factor $2/T$ is not included because it cancels out within the ratio, seen for example in equations 4.39 and 4.40.

The Butterworth nth order filter (see, for example, Proakis & Manolakis(11)) is given by

$$|H(\omega)|^2 = \frac{1}{1+(\omega/\omega_c)^{2n}} \tag{4.39}$$

where $\omega_c = \omega_{a1} = 0.325$. The order n is found from the condition for the above response to be 10 dB down to $\omega_{a2} = 0.726$, i.e.

$$1+(0.726/0.325)^{2n} = 10$$

giving $n = 1.367$, therefore we choose $n = 2$. A second-order Butterworth filter with $\omega_c = 0.325$ has poles at $s_{1,2} = 0.325(0.707 \pm j \times 0.707) = 0.23 \pm j0.23$, and no zeros. The transfer function is given by

$$H(s) = \frac{s_1 s_2}{(s+s_1)(s+s_2)} = \frac{0.1058}{s^2+0.46s+0.1058} \tag{4.40}$$

Replacing s by $(1-z^{-1})/(1+z^{-1})$ in the above expression for $H(s)$, we have

$$H(z) = \frac{0.1058}{\left(\dfrac{1-z^{-1}}{1+z^{-1}}\right)^2 + 0.46\left(\dfrac{1-z^{-1}}{1+z^{-1}}\right)+0.1058}$$

or, after rearrangement

$$H(z) = 0.0676\frac{1+2z^{-1}+z^{-2}}{1-1.141z^{-1}+0.413z^{-2}} \tag{4.41}$$

which is the transfer function of the required digital filter. The corresponding difference equation is

$$\begin{aligned} y(k) &= 0.0676x(k)+0.135x(k-1)+0.0676x(k-2) \\ &\quad +1.141y(k-1)-0.413y(k-2) \end{aligned} \tag{4.42}$$

which follows from equation 4.41.

The frequency response is given by

$$H(\omega) = H(z = e^{j\omega T}) = 0.0676\frac{1+2e^{-j\omega T}+e^{-j2\omega T}}{1-1.141e^{-j\omega T}+0.413e^{-j2\omega T}}$$

from which the magnitude squared is

$$|H(m)|^2 = 0.1352\frac{3+4\cos m\pi + \cos 2m\pi}{2.473 - 3.224\cos m\pi + 0.826\cos 2m\pi} \tag{4.43}$$

where as before $m\pi = \omega T$. The function $10\log_{10}|H/H(0)|^2$ has been calculated for $0 \le m \le 2$ in steps of 0.05, using the computer program given in section 5 of the appendix. The result is shown in fig. 4.8, where the specification points are also inserted.

The bilinear z-transform, the impulse invariance, and other analogue-to-digital filter transformations can be obtained by other approaches. One particularly interesting approach is based on the convolution integral expressed as an integro-difference

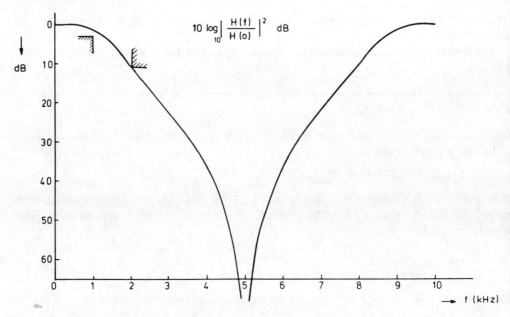

Fig. 4.8 Frequency response for digital second-order Butterworth filter

equation, and is discussed by Haykin(19). Various transformations can then be obtained, depending on the specific type of approximation of the continuous-time input signal. There are also other interpretations (for example, as described by Oppenheim & Schaffer(10), section 5), and an interesting approach is also shown in problem 4.10.

4.3   Frequency transformations of lowpass filters

The design methods discussed in the previous sections, i.e. the impulse invariance and bilinear z-transform, perform transformation of lowpass analogue filter to digital lowpass filter. These transformations operate directly on the analogue filter variable s and produce the digital filter in variable z^{-1}. Other types of filters, such as highpass, bandpass and bandstop, can also be obtained by similar transformations of the lowpass analogue filter as discussed by Bogner & Constantinides(18), chapter 4.

Table 4.1 lists various transformations, from variable s to variable z^{-1}, for a given lowpass analogue filter of cutoff frequency Ω_c to a digital lowpass or highpass filter of cutoff frequency ω_c. They also enable transformation of a given lowpass to bandpass and bandstop filters, which are specified in terms of the upper (ω_2), lower (ω_1) and centre (ω_0) frequencies. A bandpass design example, based on Table 4.1, is given next.

Example 4.5
A digital signal processing system has a sampling rate of 1 kHz. Design a digital bandpass filter for the system with the following specifications:
(a) the range of the passband is from 100 to 400 Hz with ripple-free attenuation between 0 and 3 dB;
(b) attenuation must be at least 20 dB at 45 and 450 Hz, and must fall off monotonically beyond these frequencies.

Table 4.1   Transformations from analogue lowpass filter variable s to digital filter variable

Required digital filter	Replace s by:	Associated design parameters
Lowpass	$\beta\dfrac{1-z^{-1}}{1+z^{-1}}$	$\beta = \Omega_c \cot\dfrac{\omega_c T}{2}$
Highpass	$\beta\dfrac{1+z^{-1}}{1-z^{-1}}$	$\beta = \Omega_c \tan\dfrac{\omega_c T}{2}$
		ω_c = desired cutoff frequency
Bandpass	$\beta\dfrac{z^{-2}-2\alpha z^{-1}+1}{1-z^{-2}}$	$\alpha = \cos\omega_0 T$
		$= \dfrac{\cos\dfrac{\omega_2+\omega_1}{2}T}{\cos\dfrac{\omega_2-\omega_1}{2}T}$
		$\beta = \Omega_c\cot\dfrac{\omega_2-\omega_1}{2}T$
Bandstop	$\beta\dfrac{1-z^{-2}}{z^{-2}-2\alpha z^{-1}+1}$	α = as above
		$\beta = \Omega_c\tan\dfrac{\omega_2-\omega_1}{2}T$
		$\omega_2,\ \omega_1$ = desired upper and lower cutoff frequencies
		ω_0 = centre frequency

For the bandpass filter, using the expressions given in Table 4.1, we have

$$\alpha = \cos\omega_0 T = \frac{\cos[(f_2+f_1)/f_s]\pi}{\cos[(f_2-f_1)/f_s]\pi}$$

$$\beta = \Omega_c \cot[(f_2-f_1)/f_s]\pi$$

Using values for f_1, f_2 and f_s as specified, we obtain

$$\alpha = \cos\omega_0 T = \frac{\cos \pi/2}{\cos 0.3\pi} = 0 \qquad (4.44)$$

$$\beta = \Omega_c \cot 0.3\pi = (0.72654253)\Omega_c \qquad (4.45)$$

The transformation from s to z^{-1} plane, in this case, is

$$s = \beta\frac{1+z^{-2}}{1-z^{-2}} \qquad (4.46)$$

Table 4.2   Digital bandpass and equivalent analogue lowpass frequencies

Digital bandpass	Equivalent analogue lowpass
$\omega_d = 2\pi\,45\,\text{rad s}^{-1}$	$\omega_{b1} = -2.5501\,\Omega_c$
$\quad = 2\pi\,100$	$\omega_{a1} = -\Omega_c$
$\quad = 2\pi\,400$	$\omega_{a2} = \Omega_c$
$\quad = 2\pi\,450$	$\omega_{b2} = 2.2361\,\Omega_c$

Therefore, the analogue lowpass filter and digital bandpass filter frequencies are related by

$$\omega_a = -\beta \cot \omega_d T \tag{4.47}$$

Using this relationship we obtained the frequencies given in Table 4.2. Since filter response must be monotonic, we choose the Butterworth filter for which

$$|H(\omega)|^2 = \frac{1}{1 + (\omega/\Omega_c)^{2n}} \tag{4.48}$$

where the cutoff $\Omega_c = 1$ (see for example (10), (11) and (17). The order of filter (n) is determined from the requirement that the above function is at least -20 dB at 45 and 450 Hz.

Therefore we have

$$1 + (2.23/1)^{2n} = 100$$

where $\omega_{b2} = 2.23\,\text{rad s}^{-1}$ has been taken as the more stringent requirement then $\omega_{b1} = -2.55\,\text{rad s}^{-1}$.

The solution for n is then

$$n = 1/\log_{10} 2.23 = 2.87$$

hence we choose $n = 3$. The equivalent analogue lowpass (or prototype) filter is therefore third-order Butterworth with the transfer function

$$H(s) = \frac{1}{s^3 + 2s^2 + 2s + 1} \tag{4.49}$$

The digital filter is then obtained using the substitution

$$s = 0.72654253\frac{1 + z^{-2}}{1 - z^{-2}} \tag{4.50}$$

in equation 4.49. The final expression for the digital bandpass filter transfer function is given as

$$H(z) = \frac{0.25691560(1 - 3z^{-2} + 3z^{-4} - z^{-6})}{1 - 0.57724052z^{-2} + 0.42178705z^{-4} - 0.05629724z^{-6}} \tag{4.51}$$

where the numerical values have been calculated on a pocket calculator. The filter structure can be obtained, as before in example 4.2, by expressing $H = Y/X$.

To test the frequency response of the above filter, we have used the computer program given in section 5 of the appendix to calculate the function

$$10\log_{10}|H(m)/H(0.5)|^2 \text{ dB}$$

where $|H(m)|^2$ is equal to

$$\frac{0.13201125 \times 10 - 15 \cos 2m\pi + 6 \cos 4m\pi - \cos 6m\pi}{1.51428032 - 1.68891710 \cos 2m\pi + 0.90856820 \cos 4m\pi - 0.11259448 \cos 6m\pi} \qquad (4.52)$$

for $m = 0, 0.05, 0.10, \ldots, 2$.

The result is plotted in fig. 4.9 where the specifications are also shown.

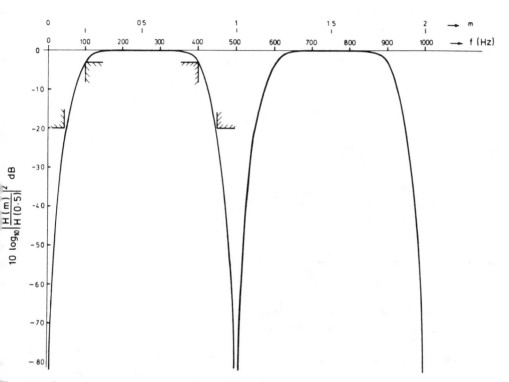

Fig. 4.9   Frequency response for digital bandpass filter

General frequency transformation schemes are indicated in fig. 4.10 with reference to table 4.1 and table 4.3. The transformations in table 4.1 link directly analogue and digital filters, but the transformations in table 4.3 relate only to digital filters, and to get to the analogue domain we have to use the bilinear z-transform marked by BT. Another possible path refers to analogue transformations not given here, but they can be found for example in Proakis & Manolakis(11).

Example 4.6

The same specification as in example 4.5, but the design uses table 4.3 and BT. From the digital BP specification we have $\omega_2 T = 0.8\pi$, $\omega_1 T = 0.2\pi$ and the bandwidth $\omega_c T = (\omega_2 - \omega_1)T = 0.6\pi$. Using these values in BP/LP transformation we find $\alpha = 0$ and $k = 1$ which produces the transformation relationship

$$Z^{-1} = -z^{-2} \text{ or } Z = -z^2 \qquad (4.53)$$

Table 4.3 Frequency transformations for digital filters

Filter type	Transformation	Parameters
Lowpass	$Z^{-1} \rightarrow \dfrac{z^{-1} - \alpha}{1 - \alpha z^{-1}}$	$\alpha = \dfrac{\sin[(\omega_c - \omega_c')T/2]}{\sin[(\omega_c + \omega_c')T/2]}$

ω_c = prototype lowpass cutoff frequency

ω_c^1 = desired cutoff frequency

| Highpass | $Z^1 \rightarrow = \dfrac{z^{-1} + \alpha}{1 + \alpha z^{-1}}$ | $\alpha = \dfrac{-\cos[(\omega_c - \omega_c')T/2]}{\cos[(\omega_c + \omega_c')T/2]}$ |

| Bandpass | $Z^{-1} \rightarrow - \dfrac{z^{-2} - \dfrac{2\alpha k}{k+1} z^{-1} + \dfrac{k-1}{k+1}}{\dfrac{k-1}{k+1} z^{-2} - \dfrac{2\alpha k}{k+1} z^{-1} + 1}$ | |

$$\alpha = \frac{\cos[(\omega_2 + \omega_1)T/2]}{\cos[(\omega_2 - \omega_1)T/2]} \quad, k = \tan(\frac{\omega_c T}{2})/\tan[(\omega_2 - \omega_1)T/2]$$

| Bandstop | $Z^{-1} \rightarrow \dfrac{z^{-2} - \dfrac{2\alpha}{1+k} z^{-1} + \dfrac{1-k}{1+k}}{\dfrac{1-k}{1+k} z^{-2} - \dfrac{2\alpha}{1+k} z^{-1} + 1}$ | |

α = as above, $k = \tan(\omega T/2) = \tan[(\omega_2 - \omega_1)T/2)]$

ω_2, ω_1 = desired upper and lower cutoff frequencies

Z^{-1} refers to the prototype lowpass filter

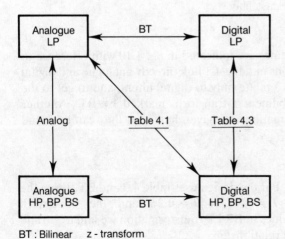

BT : Bilinear   z - transform

Fig. 4.10 Frequency transformation schemes

Converting to the frequency domain we have

$$e^{j\Omega_d T} = -e^{-2j\omega_d T} = e^{j(\pi - 2\omega_d T)}$$

Therefore $\Omega_d T = \pi - 2\omega_d T$ $\hspace{4cm}$ (4.54)

where Ω_d refers to digital LP and ω_d to digital BP filters. Using equation 4.54, BP frequencies

$$\omega_d T = \{0.9\pi, 0.8\pi, 0.2\pi, 0.09\pi\}$$

transform to LP frequencies

$$\Omega_d T = \{-0.8\pi, -0.6\pi, 0.6\pi, 0.82\pi\}$$

Next we use the bilinear transform (BT):

$$\Omega_a = \tan(\Omega_d T/2) \hspace{4cm} (4.55)$$

$$s = (Z-1)/(Z+1) \hspace{4cm} (4.56)$$

To simplify, the factor $T/2$ has been left out, since it cancels when returning to digital LP. Applying equation 4.55 to the set $\Omega_d T$, we obtain

$$\Omega_a = \{-3.077683, -1.37638, 1.37638, 3.4420\}$$

The more stringent requirement is for the first two frequencies as $3.077683/1.37638 = 2.24$. This is almost the same value as in example 4.5, hence we can use a Butterworth filter with $n = 3$ as given by equation 4.49.

The procedure is now reversed by first dividing s variable in equation 4.49 by the analogue cutoff frequency $\Omega_{ac} = 1.37638$ and substituting equation 4.56 for s, which brings us back to digital LP with the transfer function $H(z)$. Applying now the transformation given by equation 4.53 we transform the digital LP to the required BP digital filter. The resulting transfer function is the same as equation 4.52 obtained by the other transformation method in example 4.5.

Note: detailed design examples using the transformation in table 4.3 can be found in Kuc(17).

An interesting alternative approach to bandpass filter design is shown in the following example.

Example 4.7

The transfer function of a discrete time network given as

$$H(z) = \frac{a_0 z^2 + a_1 z}{z^2 + b_1 z + b_2} \hspace{4cm} (4.57)$$

is to have BP filter characteristics. Zeros of this function $z_1 = 0$, $z_2 = -a_1/a_0$, assuming $a_1 < a_0$, and poles

$$p_{1,2} = -\frac{b_1}{2} \pm j \sqrt{b_2 - \frac{b_1^2}{4}} = \sqrt{b_2} \exp\left[\pm j \cos^{-1}\left(-\frac{b_1}{2\sqrt{b_2}}\right)\right]$$

are shown in fig. 4.11. For stability poles must be inside the unit circle, hence $b_2 < 1$. As indicated in fig. 4.11, the resonance frequency is given by

$$f_0 = \frac{1}{2\pi T} \cos^{-1}(b_1/2\sqrt{b_2}) \hspace{4cm} (4.58)$$

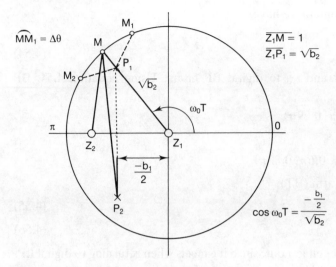

Fig. 4.11   Pole-zero pattern for example 4.7

This frequency varies with b_1 and to a lesser degree with b_2, but it is effectively controlled by varying the clock frequency, i.e. by changing T.

With reference to fig. 4.11, and section 2.4, the magnitude of the transfer function is given by

$$|H(\omega_o)| = G\frac{\overline{Z_1 M} \cdot \overline{Z_2 M}}{\overline{P_1 M} \cdot \overline{P_2 M}} \tag{4.59}$$

The length $\overline{P_1 M}$ is the most significant since it is in the denominator and it is the smallest of four distances from poles and zeros to the point M. From equation 4.59 we can deduce that for $-3\,\mathrm{dB}$ point we require $\overline{P_1 M} \to \overline{P_1 M}\sqrt{2}$, which means $\overline{P_1 M_1} = \overline{P_1 M_2} = \overline{P_1 M}\sqrt{2}$. Therefore we have

$$\overline{MM_1} \cong \Delta\theta = \sqrt{[(\overline{P_1 M_1})^2 - (\overline{P_1 M})^2]} = \overline{P_1 M} = 1 - \sqrt{b_2} \tag{4.60}$$

and forming the ratio $\Delta\theta/\pi$, where $\Delta\theta = 2\pi\Delta f T$, we obtain

$$\Delta f = (1 - \sqrt{b_2})/2\pi T \tag{4.61}$$

The Q-factor is then given by

$$Q = \frac{f_0}{2\Delta f} = \frac{\cos^{-1}(-b_1/2\sqrt{b_2})}{2(1 - \sqrt{b_2})} \tag{4.62}$$

Equations 4.58 and 4.62 enable us to determine filter parameters b_1, b_2 to obtain the resonant circuit of desired Q-factor. More detail and implementation in sampled date form can be found in Smith et al.[20].

4.4   Implementation of discrete-time systems

Some typical realization structures have been introduced in section 2.5. They consist of delay elements, multipliers and summers. Their hardware implementation can be done in analogue discrete-time form or in digital form. The discrete-time implementation of summers and multipliers is straightforward by means of the operational amplifiers, but the delay elements require special (charge-coupled) units, Bozic(21). On the other hand for digital implementation delays and adders are straightforward but the multipliers are specialized units requiring considerable attention.

In digital implementation signals and coefficients are restricted to a finite set of possible values, or quantized levels, dictated by the register size. It is then important to consider quantization effects in the fixed-point representation with numbers normalized to be less than one. We consider only briefly the effects of quantization on filter performance in the following order: quantization of coefficients (a_m, b_m), quantization of the input signal, and quantization of products of signals and coefficients.

4.4.1   Coefficient quantization

The analysis is done in the z-domain using the transfer function expressed in terms of poles and zeros (equation 2.17). Considering first the poles one can show, Proakis and Manolakis(11) that:

$$\Delta p_i = - \sum_{m=1}^{N} \frac{p_i^{N-m}}{\prod_{j=1, j \neq i}^{N}(p_i - p_j)} \Delta b_m \qquad (4.63)$$

where Δb_m is the change in b_m due to rounding or truncation necessary to fit the finite length registers, and $(p_i - p_j)$ are vectors in the z-plane from poles p_j to pole p_i. If $(p_i - p_j)$ vectors are small, the reciprocals are large and hence Δp_i due to Δb_m can be large. This will in turn change the transfer function $H(z)$ and consequently the frequency response. One can minimize this effect by maximizing $(p_i - p_j)$ by using second-order units as discussed in section 2.5. The same type of analysis and comments apply to zeros in $H(z)$. Therefore, FIR with a large number of zeros should also be realized as cascades of first- and second-order sections.

4.4.2   Input signal quantization

The general scheme is shown in fig. 4.12, where $x_a(t)$ is the input analogue signal, $x(n)$ is its sampled value, and $x_q(n)$ its quantized value. The error produced by the quantization, $e(n)$, is referred to as the quantization noise with zero mean and variance $\sigma_e^2 = q^2/12$.
    The quantization step is given by

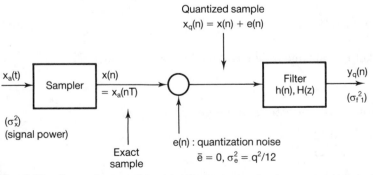

Fig. 4.12   General scheme for input/output quantization

$$q = \frac{R}{2^{b+1}} \tag{4.64}$$

where R is peak-to-peak quantizer range and $(b+1)$ is the number of bits per register (or word). The signal to quantization noise power ratio, $\text{SQNR} = 10\log_{10}(\sigma_x^2/\sigma_e^2)$, can be shown to be

$$\text{SQNR} = 6b + 16.8 - 20\log\frac{R}{\sigma_x} \tag{4.65}$$

Various SQNR expressions can be obtained depending on the type of signal applied to analogue-to-digital converter (ADC). However, the main point here is that SQNR increases by 6 dB for each additional bit in the quantizer.

The variance of the output noise, σ_{f1}^2, due to ADC is

$$\sigma_{f1}^2 = \sigma_e^2 \left[\sum_{n=-\infty}^{\infty} h^2(n) \right]$$

$$= \sigma_e^2 \left[\frac{1}{2\pi} \oint_{|z|} H(z)H(z^{-1})z^{-1}\mathrm{d}z \right] \tag{4.66}$$

where the contour integration within the square bracket is simply given as

$$\Sigma \text{ Residues } [H(z)H(z^{-1})z^{-1}] \text{ at poles inside unit circle}$$

$$= \sum_i \{(z-p_i)[H(z)H(z^{-1})z^{-1}]\}_{z=p_i} \tag{4.67}$$

Example 4.8
For $H(z) = (1-a)/(1-az^{-1})$ one can show that $\sigma_{f1}^2 = [(1-a)^2/(1-a^2)]\sigma_e^2$

4.4.3   Product quantization in filter

For the fixed point arithmetic the product of two b-bits numbers produces a $2b$-bits long result which needs to be rounded or truncated to b-bits. This operation produces the quantization noise as indicated in fig. 4.13(a).

In the second order section, fig. 4.13(b), there are four products producing noise sources e_1, e_2, e_3, e_4. The output noise variance, in this case, is given by

$$\sigma_{f2}^2 = 2\sigma_e^2 \left[\sum_{n=-\infty}^{\infty} h^2(n) \right] + 2\sigma_e^2 \tag{4.68}$$

where the square bracket is evaluated as in equation 4.66 by equation 4.67.

Example 4.9
The transfer function for a second-order section is given by

$$H(z) = \frac{1 - 0.8z^{-1} + 0.12z^{-2}}{1 - 1.2z^{-1} + 0.32z^{-2}}$$

Using equation 4.68, and (4.66/4.67), one can show that $\sigma_{f2}^2 = 3.95\sigma_e^2$

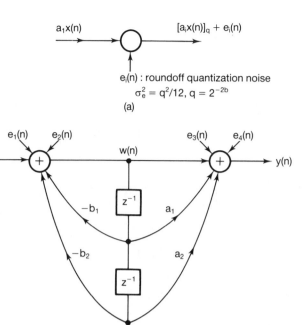

e$_i$(n) : roundoff quantization noise

$$\sigma_e^2 = q^2/12, q = 2^{-2b}$$

(a)

Fig. 4.13 (a) Product quantization method (b) second-order section with product quantization noise sources

4.4.4 Oscillations and other comments

The type of noise discussed so far is the dominant component of the output noise in a digital filter only when the addition overflows in the internal registers are negligible. Therefore, it is important to scale the unit sample response between the input and any internal summing nodes. If not controlled the overflow can cause large oscillations due to the two's complement arithmetic system normally used in digital filters.

Another type of oscillation known as limit cycles can also occur in recursive systems due to round-off errors in multiplication.

More detail on all of the material in this section can be found in Proakis & Manolakis(11) and deFatta(22).

4.5 Wave digital filter (WDF)

They are a special class of digital filters based on the classical LC-ladder filters. The low sensitivity of LC-ladders to the component changes is transferred to the WDF, hence they can be realized with a smaller number of bits. The link between them is established in terms of the transmission line theory (incident and reflected waves).

We introduce the concepts by considering the third order LC ladder filter drawn in a special way, fig. 4.14. The source, load and filter elements (C_1, L_2, C_3) are shown as 1-ports, while their interconnections are 3-ports (two parallel and one serial type, in this case). Each port is then described in terms of the incident wave (A), reflected wave (B), and port resistance (R) as follows:

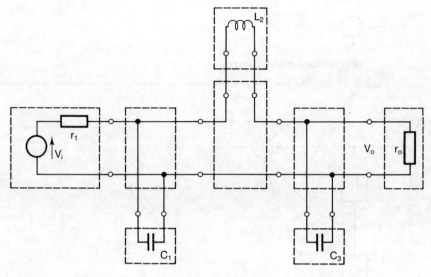

Fig. 4.14   Third-order analogue ladder prepared for WDF transformation

$$A = V + RI \atop B = V - RI \Big\}$$

(4.69)

The derivations for 1-ports are straightforward, but for junction 3-ports they are more complicated, Antoniou(23), Bozic(24). The resulting WDF structure is shown in fig. 4.15. It is seen that the 1-port units for the input and output are simply $B_i = V_i$, and $A_0 = 2V_0$, capacitors are represented by delays and inductors by delay and an inverter. The junctions are shown by two parallel adaptors (P1, P2) and a serial adaptor (S1). Their incident and reflected waves are related, in this case, in terms of 3×3 matrices. The symbols used for the adaptors are the ones normally used in the literature. These symbols look deceptively simple but, in fact, the adaptors are the main complexity for a

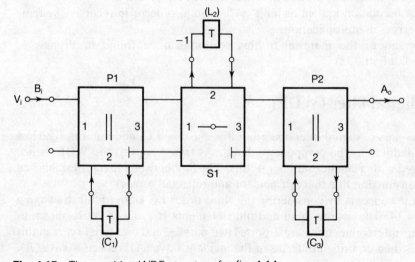

Fig. 4.15   The resulting WDF structure for fig. 4.14

hardware implementation. Their internal structure contains the following elements: P1 (4 adders, 1 multiplier and 1 inverter); S1 (4 adders, 1 multiplier and 3 inverters); P2 (6 adders, 2 multipliers and 1 inverter).

The WDF realization in fig. 4.15 can be greatly simplified if or when the adaptors become available in microcircuit form. Then, it should be quite straightforward to convert an LC-type ladder filter, with its low sensitivty, into the corresponding WDF structure. On the other hand, one can realize the WDFs using special-purpose hardware controlled by a microprocessor, Tan(25).

4.6 Problems

4.1 An analogue filter is specified in terms of a Butterworth second-order filter with the transfer function

$$H(s) = \frac{1}{s^2 + s\sqrt{2} + 1}$$

where the cutoff if $\Omega_c = 1$ rad s^{-1}.
(a) Derive a single digital filter transfer function using the impulse invariance method.
(b) From the result in (a) determine the filter structure in terms of the difference equation.

4.2 (a) For the analogue function given in problem 4.1, derive a single digital filter using the bilinear z-transform method.
(b) Determine the difference equation for this filter and compare it with the difference equation of problem 4.1 for $T \to 0$ (very frequent sampling).

4.3 A first-order Butterworth filter normalized to $\Omega_c = 1$ rad s^{-1}, is given by

$$H(s) = \frac{1}{s + 1}$$

To normalize it to the cutoff frequency ω_c, we have to replace s by s/ω_c.
(a) Show how the corresponding impulse invariant digital filter is affected by the change of s to s/ω_c.
(b) How can the solution of problem 4.1 be changed to make it valid for the cutoff ω_c?

4.4 Design a lowpass digital filter with cutoff frequency ω_c, based on the analogue lowpass filter

$$H(s) = \frac{1}{s^2 + s\sqrt{2} + 1}$$

which is normalized to $\Omega_c = 1$ rad s^{-1}.
(a) Use Table 4.1 to find the solution in terms of β.
(b) Assuming that $\omega_c T/2$ is small so that $1/\beta \simeq \omega_c T/2$, derive the difference equation.
(c) Compare this solution with the one of problem 4.2. Note how ω_c enters the solution in this case.

4.5   (a) Design a single digital highpass filter of cutoff frequency ω_c based on the analogue lowpass filter as specified in problem 4.4. Compare the digital transfer function obtained for this filter with the one in problem 4.4. Can the highpass be obtained from the lowpass digital filter transfer in a simple manner?
(b) Derive the highpass filter structure for the sampling frequency $f_s = 4f_c$. Note: It is advantageous to carry out the analysis in terms of β, and substitute for its value only in the last stages of the exercise.

4.6   Design a digital bandpass filter based on the lowpass analogue filter specified in problem 4.1. The passband range is to be from 100 to 700 Hz, and the system sampling frequency is 2.4 kHz.

4.7   Design a digital bandstop filter based on the lowpass analogue filter of problem 4.1. The stopband range is to be from 100 to 700 Hz, and the sampling frequency is 2.4 kHz.

4.8   The digital filters described by equations 4.16 and 4.37 are obtained respectively by application of the impulse invariant and bilinear z-transform methods to the first-order RC filter. Show that these two filters reduce to the same difference equation if sampling frequency is very large, in which case $\omega_c T \ll 1$.

4.9   (a) Apply the bilinear z-transform to the first-order analogue filter

$$H(s) = \frac{a}{s+a}$$

and show that the digital and analogue frequencies are related by equation 4.30.
(b) The impulse invariant transform of the above analogue filter is given by equation 4.11. Form the ratio $|H(\omega)/H(0)|$ for both the analogue and digital filter. Show that the analogue and digital frequencies are, in this case, related by

$$\omega_a = a\frac{\sin(\omega_d T/2)}{\sinh(aT/2)}$$

(c) Show that for very frequent sampling, i.e. $T \to 0$, both results (a) and (b) reduce to $\omega_a = \omega_d$.

4.10 The Laplace and z-transform variables are related by $z = e^{sT}$ which can be written as

$$s = \frac{1}{T}\ln z$$

The $\ln z$ function can be approximated in the following three ways:

$$AP(1) = (z-1) - \tfrac{1}{2}(z-1)^2 + \tfrac{1}{3}(z-1)^3 - \dots$$

$$AP(2) = \left(\frac{z-1}{z}\right) + \tfrac{1}{2}\left(\frac{z-1}{z}\right)^2 + \tfrac{1}{3}\left(\frac{z-1}{z}\right)^3 + \dots$$

$$AP(3) = 2\left[\left(\frac{z-1}{z+1}\right) + \tfrac{1}{3}\left(\frac{z-1}{z+1}\right)^3 + \tfrac{1}{5}\left(\frac{z-1}{z+1}\right)^5 + \dots\right]$$

see, for example, Abramowitz *et al.*(26), p. 68.

The exact value for $\ln(z = e^{j\omega T})$ is $j2\pi f/f_s$, where $f_s = 1/T$ is the sampling frequency. Assuming the sampling frequency $f_s = \lambda f$, where f is the highest frequency of interest in a given system, then we have $\ln z = j2\pi/\lambda$, and we define the error as

$$e_N = \left| \frac{\ln z - AP(N)}{\ln z} \right| \times 100\%$$

where $AP(N)$ represents the three approximations listed above. Taking only the first term in each $AP(N)$, calculate e_N, for $N = 1$, 2, 3, if $\lambda_1 = 2\pi \times 100$ and $\lambda_2 = 2\pi \times 10$. Note that the first term of $AP(3)$, i.e. $2[(z-1)/(z+1)]$, is the bilinear z-transform.

5

Further concepts in discrete-time filtering

5.0 Introduction

In the first chapter we dealt with finite and infinite length discrete-time sequences, and introduced their z-transform representation. From this, the frequency spectrum has been obtained by the substitution $z = e^{j\omega T}$ as discussed in section 2.3.

In this chapter we return to the basic concepts, considering first a periodic discrete-time sequence and derive its discrete Fourier series (DFS). Having established these relationships we show how the frequency spectrum of a finite (aperiodic) sequence is related to the periodic sequence frequency spectrum, and so obtain the discrete Fourier transform (DFT).

The other feature presented in this chapter is the inverse filter which shapes a finite input sequence into a unit pulse, or 'spike'. The least output error energy is then introduced as a criterion for derivation of the optimum finite length filter. This analysis is conducted in the time-domain which is convenient for solving some problems, for example in communications, radar, sonar and seismic explorations.

5.1 Derivation of discrete Fourier series (DFS)

This section uses some results derived in section 1.4. We established there two forms of frequency spectrum $F_s(s)$ for a sampled signal, expressed in terms of equations 1.17 and 1.20. Writing these two equations for $s = j\omega$, and equating them, we have

$$\sum_{n=-\infty}^{\infty} F(\omega + n\omega_s) = T \sum_{k=0}^{\infty} f(kT) e^{-jk\omega T} \tag{5.1}$$

The left-hand side represents the periodic spectrum discussed in section 1.4. Denoting it here as $F_p(\omega)$ we obtain

$$F_p(\omega) = T \sum_{k=0}^{\infty} f(kT) e^{-jk\omega T} \tag{5.2}$$

For illustration we consider, in example 5.1, a simple discrete-time sequence.

Example 5.1
The sequence $f(kT) = e^{-akT}$, for $k > 0$, is shown in fig. 5.1(a) assuming $aT = 1$. It decays fast, therefore we can take for the upper limit $k = 5$, and calculate its frequency spectrum from equation 5.2 as

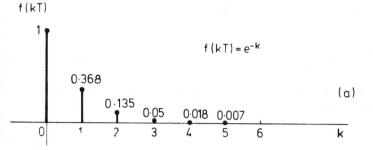

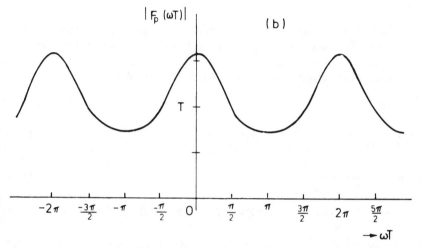

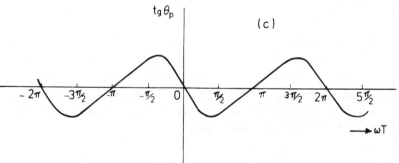

Fig. 5.1   (a) Decreasing time sequence; (b) magnitude and (c) phase of frequency spectrum of (a)

$$F_p(\omega) = T \sum_{k=0}^{5} e^{-kaT} e^{-jk\omega T}$$

or $$F_p(\omega) = T(1 + e^{-aT}e^{-j\omega T} + \ldots + e^{-5aT}e^{-j5\omega T})$$

which is a finite geometric series. Summing this series gives

$$F_p(\omega) = T\frac{1 - e^{-6aT}e^{-j6\omega T}}{1 - e^{-aT}e^{-j\omega T}} \qquad (5.3)$$

For $aT = 1$, $e^{-6aT} = e^{-6} = 0.0025$, and hence the numerator can be taken as unity, because also $|e^{-j6\omega T}| = 1$, so

$$F_p(\omega) = \frac{T}{1 - e^{-aT}e^{-j\omega T}} \tag{5.4}$$

It is interesting at this point to check this result for $T \to 0$, i.e. the case of very frequent sampling. We can then approximate

$$e^{-(a+j\omega)T} \simeq 1 - (a + j\omega)T$$

giving $F(\omega) = \dfrac{1}{a + j\omega}$

which is the familiar Fourier spectrum (aperiodic) for the exponential function $f(t) = e^{-at}$ where $t > 0$. This also shows that the spectrum relationships established earlier, equations 5.1 and 5.2, are correct.

The sampled-data spectrum given in equation 5.4 can be written as

$$F_p(\omega) = |F_p(\omega)| e^{j\theta_p}$$

where $|F_p(\omega T)| = \dfrac{T}{(1 - 2e^{-aT}\cos\omega T + e^{-2aT})^{1/2}}$ $\tag{5.5}$

and $\tan\theta_p = \dfrac{e^{-aT}\sin\omega T}{1 - e^{-aT}\cos\omega T}$ $\tag{5.6}$

Both of these functions are periodic in frequency, as shown in fig. 5.1(b) and (c), for $aT = 1$.

Returning now to equation 5.2 and taking a finite length sequence with upper limit $(N-1)$, we have

$$F_p(\omega) = T \sum_{k=0}^{N-1} f(kT) e^{-jk\omega T} \tag{5.7}$$

Assuming a periodic sequence $f_p(kT)$ with period $T_p (= NT)$ as shown in fig. 5.2, the frequency spectrum, which is already periodic because $f(t)$ is sampled, becomes discrete in frequency because $f(t)$ is periodic. We denote the discrete frequency spacing by $\Omega = 2\pi/T_p$, where T_p is the period of the time sequence.

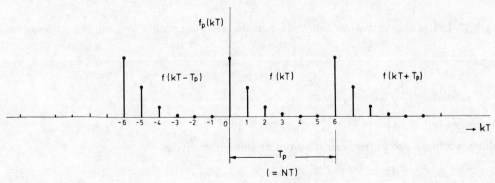

Fig. 5.2   Periodic sequence

The spectrum, equation 5.7 is now rewritten as

$$F_p(m\Omega) = T \sum_{k=0}^{N-1} f_p(kT) e^{-jkm\Omega T} \tag{5.8}$$

The next step is to derive an expression for $f_p(kT)$ in terms of the spectrum $F_p(m\Omega)$. For this purpose we multiply the above equation by $e^{jmn\Omega T}$ and sum over m as follows:

$$\sum_{m=0}^{N-1} F_p(m\Omega) e^{jmn\Omega T} = T \sum_{m=0}^{N-1} \sum_{k=0}^{N-1} f_p(kT) e^{-jkm\Omega T} e^{jmn\Omega T}$$

$$= T \sum_{k=0}^{N-1} f_p(kT) \sum_{m=0}^{N-1} e^{j\Omega Tm(n-k)} \tag{5.9}$$

where we have taken the same number of samples (N) both in the time- and frequency-domain. Since $T_p = NT$, we have

$$\Omega = 2\pi/T_p = 2\pi/NT \tag{5.10}$$

and the second sum of equation 5.9 becomes

$$\sum_{m=0}^{N-1} e^{j(2\pi/N)m(n-k)} = \frac{1 - e^{j2\pi(n-k)}}{1 - e^{j(2\pi/N)(n-k)}} \tag{5.11}$$

which results in

$$\sum_{m=0}^{N-1} e^{j(2\pi/N)m(n-k)} = \begin{cases} N & n = k \\ 0 & n \neq k \end{cases} \tag{5.12}$$

where the first result follows from direct summation of the left-hand-side of equation 5.11, but the result for $n \neq k$ follows from the right-hand-side of equation 5.11 in which the numerator is then zero because $\exp[j2\pi(\text{integer})] = 1$. Returning to equation 5.9, we have

$$NT \sum_{k=0}^{N-1} f_p(kT) = \sum_{m=0}^{N-1} F_p(m\Omega) e^{j(2\pi/N)mk}$$

Therefore, the time sample at time kT is given by

$$NTf_p(kT) = \sum_{m=0}^{N-1} F_p(m\Omega) e^{j(2\pi/N)mk}$$

This result and equation 5.8, together with equation 5.10, form the following pair of equations:

$$F_p(m\Omega) = T \sum_{k=0}^{N-1} f_p(kT) e^{-j(2\pi/N)mk} \tag{5.13}$$

$$f_p(kT) = \frac{1}{NT} \sum_{m=0}^{N-1} F_p(m\Omega) e^{j(2\pi/N)mk} \tag{5.14}$$

to which we shall refer as the discrete Fourier series (DFS) for the periodic time sequence.

Although formulated for the periodic time-domain sequence, it will be seen in the next section that the same applies to finite (aperiodic) sequences.

5.2 Finite duration sequences – discrete Fourier transform (DFT)

We have already met the expression

$$F_p(\omega) = \sum_{n=-\infty}^{\infty} F(\omega + n\omega_s) \tag{5.15}$$

in section 1.4 and in equation 5.1 of the section 5.1. This equation shows the periodic nature of the frequency spectrum produced by sampling of a time-domain function. If the time function is periodic, then the frequency spectrum becomes discrete too. In this case, introducing $\omega = m\Omega$ and $\omega_s = N\Omega$ as discussed in the previous section, equation 5.15 can be rewritten as

$$F_p(m\Omega) = \sum_{n=-\infty}^{\infty} F(m\Omega + nN\Omega) \tag{5.16}$$

Such a function is illustrated in fig. 5.3 (for $n = -1, 0, 1$), where we have assumed that

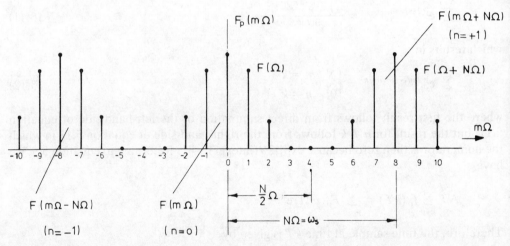

Fig. 5.3 Discrete frequency spectrum for a sampled function, periodic in time

the highest frequency in $F(m\Omega)$ is below $N\Omega/2(=\omega_s/2)$ so that the periodic sections do not overlap. Periodic spectra of this kind are of special interest since they are separable, as discussed in section 1.4.

Analogous to the above, a periodic and sampled time function can be expressed as

$$f_p(kT) = \sum_{n=-\infty}^{\infty} f(kT + nNT) \tag{5.17}$$

where T is the discrete-time interval. Such a function, without overlap between successive periods, has been shown earlier in fig. 5.2. In such cases the periodic time function of equation 5.17 is assumed to be constructed from finite duration sequences repeated at intervals of NT. Conversely, in equations 5.13 and 5.14, $f_p(kT)$ represents values of the sequence within one period. We therefore deduce that the results for DFS, given by equations 5.13 and 5.14, are also applicable to the finite duration sequences, in which case they are written as

$$F(m\Omega) = T \sum_{k=0}^{N-1} f(kT) e^{-j(2\pi/N)mk} \quad \text{where } 0 < m < N-1 \tag{5.18}$$

$$f(kT) = \frac{1}{NT} \sum_{m=0}^{N-1} F(m\Omega) e^{j(2\pi/N)mk} \quad \text{where } 0 < k < N-1 \tag{5.19}$$

The first equation is known as the discrete Fourier transform (DFT) and the second one as its inverse (IDFT). Note that although $f(kT)$ is a finite sequence, $F(m\Omega)$ is periodic, because $f(kT)$ is sampled while $F(m\Omega)$ is taken over one period.

We can also arrive at equation 5.18 by considering the z-transform of a finite duration sequence, i.e.

$$F(z) = \sum_{k=0}^{N-1} f(kT) z^{-k} \tag{5.20}$$

The sampled frequency response is then obtained by setting

$$z = e^{j\omega T} \Big|_{\omega = m\Omega} = e^{jm\Omega T} \tag{5.21}$$

and using $\Omega T = 2\pi/N$ from equation 5.10, we can write equation 5.20 as

$$F(m\Omega) = \sum_{k=0}^{N-1} f(kT) e^{-j(2\pi/N)mk} \tag{5.22}$$

which is the same as equation 5.18. The factor T is missing in the above equation, because the z-transforms given in equations 5.20 and 1.9 have been defined without T. This factor and similarly the spectral spacings Ω are often taken as unity in DFT and other similar expressions.

The computational aspects of DFT, and DFS, are illustrated in example 5.2.

Example 5.2
This example is an inversion of example 5.1, and it shows the main steps involved in calculating the time sequence from frequency sequence values. For six samples, i.e. $N = 6$, equation 5.14 gives

$$f_{\mathrm{p}}(kT) = \frac{1}{6T} \sum_{m=0}^{5} F_{\mathrm{p}}(m\Omega) e^{j(\pi/3)mk} \tag{5.23}$$

Using the expression for $F_{\mathrm{p}}(m\Omega)$, derived as equation 5.4 in example 5.1, we have

$$F_{\mathrm{p}}(m\Omega) = \frac{T}{1 - e^{-1} e^{-jm\pi/3}} \tag{5.24}$$

where we have taken $aT = 1$. It is convenient, for computational reasons, to represent the factor $e^{-jm\pi/3}$, for $m = 0$ to 5, on the unit circle diagram shown in fig. 5.4.

The time sample at $k = 0$ is given by

$$f_{\mathrm{p}}(0) = \frac{1}{6T} \sum_{m=0}^{5} F_{\mathrm{p}}(m\Omega)$$

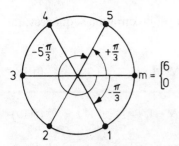

Fig. 5.4 Representation of $\exp(-jm\pi/3)$ on unit circle

Using $F_p(m\Omega)$ from equation 5.24, we have

$$f_p(0) = \frac{1}{6}\left(\frac{1}{1-e^{-1}} + \frac{1}{1-e^{-1}e^{-j\pi/3}} + \frac{1}{1-e^{-1}e^{-j2\pi/3}}\right.$$
$$\left. + \frac{1}{1+e^{-1}} + \frac{1}{1-e^{-1}e^{-j4\pi/3}} + \frac{1}{1-e^{-1}e^{-j5\pi/3}}\right)$$

With reference to fig. 5.4, we see that terms one and five can be combined, and similarly terms two and four. The result of these operations is

$$f_p(0) = \frac{1}{3}\left(\frac{1}{1-e^{-2}} + 2\frac{1-e^{-2}\cos(2\pi/3)}{1-2e^{-2}\cos(2\pi/3)+e^{-4}}\right) \tag{5.25}$$

which produces the value $f_p(0) = 1.09$. For the calculations, tables of exponential functions and a slide rule have been used. The actual value for $f_p(0)$ is unity, and the 'overshoot' is probably due to the discontinuity at the origin known as Gibbs' phenomenon. To check the method and accuracy of calculations, the sample $f_p(2)$ has also been calculated using the same procedure.

The time sample at $k = 2$ is given by

$$f_p(2) = \frac{1}{6T}\sum_{m=0}^{5} F_p(m\Omega)e^{j(2\pi/3)m}$$

The computational steps are similar to the ones used for $f_p(0)$, but now we have to take into account the factor $e^{j(2\pi/3)m}$, and diagrams similar to fig. 5.4 are very useful. The result is

$$f_p(2) = \frac{1}{3}\left(\frac{1}{1-e^{-2}} + \frac{e^{-1}-\cos(\pi/3)}{1-2e^{-1}\cos(\pi/3)+e^{-2}} - \frac{e^{-1}+\cos(\pi/3)}{1+2e^{-1}\cos(\pi/3)+e^{-2}}\right) \tag{5.26}$$

with the value $f_p(2) = 0.135$, which is in good agreement with the actual value of this time function as shown in fig. 5.1(a). We note that the time sequence in example 5.1 is taken as a finite duration sequence, but the time sequence components in example 5.2 are assumed to belong to a periodic time sequence as discussed earlier.

The above example is relatively simple but in general the numerical calculations are quite laborious, particularly if the number of samples N increases. The number of multiplications is N^2, i.e. N multiplications for each of the N frequency points. If N is a power of two, we need only $N\log_2 N$ multiplications. This efficient computational procedure is known as the fast Fourier transform (FFT), and it has made the DFT practical in many applications. For example if $N = 64 = 2^6$, we need only $N \log_2 N = 384$ multiplications, instead of $N^2 = 4096$. More detailed discussion of DFS, DFT, and FFT can be found in a number of books, for example, references (10), (11), (13).

5.3 Circular convolution

Given the FIR filter impulse response, $h(n)$, of length M, and the input signal, $x(n)$, of length L, the filtered output is obtained by their convolution

$$y(k) = \sum_{i=0}^{M-1} h(i)x(k-i) \tag{5.27}$$

The resulting sequence for $y(k)$ will be of length $(L + M - 1)$. This filtering operation can be performed by means of DFT. As indicated in problem 1.9, z-transform converts a convolutional operation into a multiplicative one, and the same applies to the Fourier transforms. Therefore, the DFT of equation 5.27 produces $Y(m) = H(m)x(m)$ where $0 \le m \le N-1$. The values of $y(k)$ are then computed from the inverse DFT of $Y(m)$. This method may be the more efficient in obtaining $y(k)$ than equation 5.27, if DFT and IDFT are implemented by FFT.

It can be shown, Proakis and Manolakis(11), that the multiplication of the DFTs of two sequences is equivalent to

$$y(k) = \sum_{i=0}^{N-1} h(i)x(k-i, \bmod N) \quad k = 0, 1, \ldots, N-1 \tag{5.28}$$

which is called the circular convolution, while equation 5.27 is called the linear convolution. Filtering by the circular convolution involves the same four steps as the linear convolution: folding, shifting, multiplying and summing. However, $x(k-i)$ involves time values outside the range $0 \le k \le N-1$. Therefore, the modulo N indexing mechanism has to be introduced in order to refer to the sequence elements within the periodic range. Since the linear convolution has the length $L + M - 1$, the DFT has to be of size $N \ge L + M - 1$, in order to obtain the correct sequence for $y(k)$.

Example 5.3
For $h(k) = \{1, 2, 3\}$ and $x(k) = \{1, 2, 2, 1\}$, equation 5.27 produces the result $y(k) = \{1, 4, 9, 11, 8, 3\}$. One can obtain the same result by using DFT of size $N = 6$, but it is simpler to use 8-point DFT. Then equation 5.18 for $T = 1$ gives us:

$$X(m) = \sum_{k=0}^{7} x(k)e^{-j2\pi km/8} \quad m = 0, 1, \ldots, 7 \tag{5.29}$$

$$H(m) = \sum_{k=0}^{7} h(k)e^{-j2\pi km/8} \quad m = 0, 1, \ldots, 7 \tag{5.30}$$

The 8-point IDFT, based on equation 5.19 with $T = 1$, gives

$$y(k) = \frac{1}{8} \sum_{m=0}^{7} Y(m)e^{j2\pi km/8} \quad k = 0, 1, \ldots, 7 \tag{5.31}$$

with the result $y(k) = \{1, 4, 9, 11, 8, 3, 0, 0\}$.

Note: The above method is suitable for computer use, but it is quicker for hand calculations to use equation 5.28. An explanation for modulo operation is: $[p, \bmod N] = p + qN$, for some integer q such that $0 \le [p \bmod N] < N$. For example, if $x(k) = \delta(k-3)$, then $x(k-6, \bmod 8) = \delta(k-9, \bmod 8) = \delta(k-9+8) = \delta(k-1)$, with $q = 1$. The circular convolution (equation 5.28) can be converted into a circulant matrix form which is easier to handle than equation 5.28. More information can be found in DeFatta(22).

5.3.1   Filtering of long data sequences

The input sequence $x(n)$ is often very long, especially in some real-time signal processing. Filtering by means of the DFT involves operations on a block of data, limited by the size of memory. Therefore, a long input sequence must be segmented to fixed size blocks before processing.

   Because the filtering is linear, successive blocks may be processed one at a time using DFT, and the output blocks are fitted together to form the overall output signal sequence. Resulting output is then identical to the sequence obtained if the long sequence had been processed by the linear convolution. For this purpose two methods have been developed known as the overlap-save and the overlap-add. Information about these methods can be found, for example, in Proakis and Manolakis(11) and DeFatta(22).

5.4   Inverse filter

The process of inverse filtering or deconvolution is very useful in radar, sonar and seismic problems for removing undesirable components in a time series. These components may be caused by the medium through which the signal is passing, as illustrated in fig. 5.5,

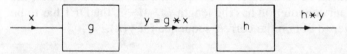

Fig. 5.5   Undesirable convolution by network g followed by deconvolution network h

where g represents the impulse response of some propagation medium causing undesirable filtering or convolution described by $g*x$. The asterisk ($*$) denotes the convolution operation as given in equation 1.6.

   The output of the following filter h is given by

$$h*y = h*g*x \tag{5.32}$$

We want to recover the signal x at the output of filter h. To achieve this we must find the filter h such that

$$h*g = \delta \tag{5.33}$$

where δ is the unit pulse sequence $[1, 0, 0, \ldots]$. Then the output of filter h is

$$h*y = \delta*x = x \tag{5.34}$$

i.e. the original signal x. It is seen that $\delta * x = x$ by applying z-transform to both sides of this equation.

To find the inverse filter we apply z-transform to equation 5.33 and obtain

$$H(z)G(z) = 1 \qquad (5.35)$$

as in, for example, problem 1.9. The required filter is therefore

$$H(z) = 1/G(z) \qquad (5.36)$$

i.e. the required filter $H(z)$ is the inverse of the signal disturbing filter $G(z)$, sometimes denoted as $h = g^{-1}$. Assuming the unit pulse response of the disturbing filter is of arbitrary but finite length, we have

$$G(z) = g(0) + g(1)z^{-1} + \ldots + g(n)z^{-n} \qquad (5.37)$$

The inverse filter of equation 5.36 is then given by

$$H(z) = \frac{1}{g(0) + g(1)z^{-1} + \ldots + g(n)z^{-n}}$$

$$= h(0) + h(1)z^{-1} + h(2)z^{-2} + \ldots \qquad (5.38)$$

where the h coefficients are obtained by long division. Therefore, the nonrecursive form of the inverse filter is of infinite length.

Example 5.4
Consider a special case of the disturbing filter described by $G(z) = 1 + az^{-1}$, i.e. $g(0) = 1$ and $g(1) = a$. The inverse filter in this case is given by

$$H(z)\frac{1}{1 + az^{-1}} = \frac{z}{z + a}$$

$$= 1 - az^{-1} + a^2 z^{-2} - a^3 z^{-3} + \ldots \qquad (5.39)$$

which is obtained by the long division. The inverse (nonrecursive) filter coefficients are: $h(0) = 1$, $h(1) = -a$, $h(2) = a^2$, $\ldots$, $h(k) = (-a)^k$. If $a < 1$, these coefficients are decreasing, but theoretically this is an infinite series solution.

The above solution is checked graphically by means of the convolution relationship $g * h = \delta$ as shown in fig. 5.6. The input sequence $g(k)$ to filter h is regarded as a sum of pulse sequences, one, $g(0) = 1$ applied at $k = 0$, and another, $g(1) = a$ applied at $k = 1$. The responses to these two samples are shown in fig. 5.6, where the final output $\delta(k)$ is obtained by addition of the individual responses. Note that this system is linear and hence the principle of superposition holds.

In practice we require finite length filters. Their output will be in error when compared to the desired output obtainable from an infinite length filter, and an analysis of finite length solutions and minimization of the error they cause are the subject of the following section.

5.5   Optimum finite inverse filter

The inverse filters introduced in the previous section are exact inverses, but they are infinitely long. In practice filters must have finite length, so they are only approximate

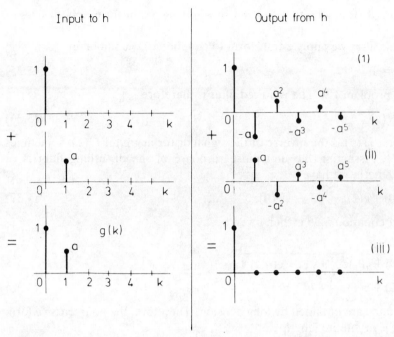

Fig. 5.6   (i) Response to 1; (ii) response to a; (iii) resultant response $= \delta(k)$

inverses whose outputs will be in error. To develop the concept of error for finite or truncated filters we use the result of example 5.4, i.e.

$$g^{-1} = h = [h(0), h(1), h(2), \ldots]$$
$$= [1, -a, a^2, -a^3, \ldots] \tag{5.40}$$

Consider first the one-length truncated inverse, i.e. $h = (1)$. Convolving this approximate inverse with the input g, we obtain the actual output λ given by

$$\lambda = g * h = (1, a) * (1) = (1, a)$$

The desired output is the pulse ('spike') at time $k = 0$, represented by $\delta = (1, 0)$, so that the error between the desired and actual output is

$$e = \delta - \lambda = (1, 0) - (1, a) = (0, -a)$$

The sum of squares of the coefficients of the error sequence represents the error energy, in this case given by

$$0^2 + (-a)^2 = a^2$$

Consider now the two-length truncated inverse, i.e. $h = (1, -a)$. The actual output is given by

$$\lambda = g * h = (1, a) * (1, -a) = (1, 0, -a^2)$$

which is easily obtained, for example, using the graphical method shown in fig. 5.6. The desired output is the pulse at time 0, i.e. $\delta = (1, 0, 0)$, and the error is

$$e = \delta - \lambda = (1, 0, 0) - (1, 0, -a^2) = (0, 0, a^2)$$

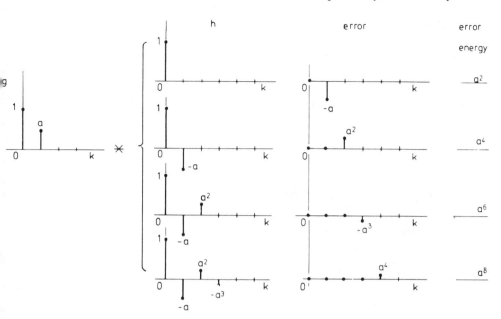

Fig. 5.7   Errors for one- to four-length truncated inverse filters

and its energy is a^4. This error energy is smaller than that for the one-length truncated inverse.

Results for one- to four-length truncated inverses are shown in fig. 5.7. We see that the error and its energy decrease as the length of the truncated (or approximate) inverse increases. However, it is possible to find an approximate inverse of a given length with smaller error energy than that of a truncated infinite inverse of the same length. To show this we derive one- and two-length inverses using the criterion of minimum error energy.

Consider first the one-length inverse $h = (h_0)$. In the truncated inverse case $h_0 = 1$, but now h_0 must be a parameter to optimize the filter in the sense of minimum error energy. The actual output is given by

$$\lambda = g * h = (1, a) * (h_0) = (h_0, ah_0)$$

and the desired output is $\delta = (1, 0)$. Therefore, the error is

$$e = \delta - \lambda = (1, 0) - (h_0, ah_0) = (1 - h_0, -ah_0)$$

with energy

$$I = (1 - h_0)^2 + (-ah_0)^2$$
$$= 1 - 2h_0 + (1 + a^2)h_0^2$$

We now minimize the above error energy with respect to h_0, i.e.

$$\frac{\partial I}{\partial h_0} = -2 + 2(1 + a^2)h_0 = 0$$

which produces the solution

$$h_0 = \frac{1}{1 + a^2} \tag{5.41}$$

for the one-length approximate inverse $h = (h_0)$.

We next consider the two-length inverse $h = (h_0, h_1)$, where again h_0 and h_1 are to be determined from the minimization of the error energy. The actual output in this case is

$$\lambda = g * h = (1, a) * (h_0, h_1) = (h_0, ah_0 + h_1, ah_1)$$

and the desired output is $\delta = (1, 0, 0)$. The error is

$$e = \delta - \lambda = (1, 0, 0) - (h_0, ah_0 + h_1, ah_1)$$
$$= (1 - h_0, -ah_0 - h_1, -ah_1)$$

with the energy given by

$$I = (1 - h_0)^2 + (ah_0 + h_1)^2 + (ah_1)^2$$
$$= 1 - 2h_0 + (1 + a^2)h_0^2 + 2ah_0h_1 + (1 + a^2)h_1^2$$

To find the optimum values for h_0 and h_1 we minimize the above expression as follows:

$$\frac{\partial I}{\partial h_0} = -2 + 2(1 + a^2)h_0 + 2ah_1 = 0$$

$$\frac{\partial I}{\partial h_1} = 2ah_0 + 2(1 + a^2)h_1 = 0$$

(5.42)

from which the optimum filter coefficients are

$$h_0 = \frac{1 + a^2}{1 + a^2 + a^4}$$

(5.43)

and $\quad h_1 = \dfrac{-a}{1 + a^2 + a^4}$

(5.44)

The corresponding minimum energy is

$$I_{min} = \frac{a^4}{1 + a^2 + a^4}$$

(5.45)

which is smaller than the value a^4 obtained earlier by truncating the exact inverse to a two-length filter. Approximate inverses of greater length can be found by the method illustrated above for one- and two-length inverses. The optimum inverses so obtained are called the least error energy approximate inverses.

The least mean error energy (or least mean-square) criteria is used in many situations (see, for example, Robinson(8) and Schwartz & Shaw(27)). We shall be using this approach again in part 2 of the book. However, while we have dealt here with deterministic signals, we shall be dealing in part 2 with random signals, and the error energies (or squares) become the mean error energies (or squares).

5.6 Multirate digital filters

All signals in previous chapters have been sampled at the same rate. However, the sampling frequency f_s can be freely chosen, but it must be larger than twice the highest frequency in the continuous signal. In many applications one needs to change the sampling rate, decreasing or increasing it. For example in communication systems operating different types of signals (speech, video, teletext, etc.) there is a requirement to process signals at different rates. Another example are the FIR filters with the implementation algorithm

$$y(n) = \sum_{i=0}^{M-1} h(n)x(n-i) \qquad (5.46)$$

which shows that they require MF multiplications per second (MPS), $(M-1)F$ additions per second (APS), where, to simplify, the sampling frequency is denoted by F. In such cases, the idea of a multirate system is to decrease the sampling rate, carry out the filtering operation at a reduced number of MPS and APS, and then increase the sampling rate back as illustrated in fig. 5.8. The decreasing sampling data rate is called decimation

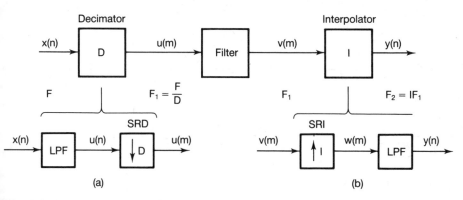

Fig. 5.8 Multirate filtering conceptual blocks

(or down sampling) by a factor D, and increasing the sampling rate by a factor I is called interpolation. By making $D = I$, the input and output rates will be the same, and the system appears externally as a single rate system.

The decimator consists of a sampling rate decreaser (SRD) which simply ignores $(D-1)$ samples out of every D input samples. This is easy to implement but care must be taken to avoid aliasing as illustrated in fig. 5.9. Aliasing can be prevented if the bandwidth of the original signal is such that $(F/D) - B \geq B$. This can be accomplished by preceding the SRD with a lowpass filter with the cutoff frequency less than or equal to $F/2D$, fig. 5.8(a). This filter can be considered as a digital anti-aliasing filter since it is

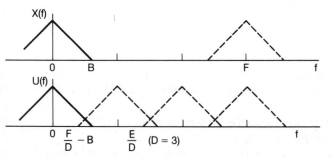

Fig. 5.9 Decimator spectra

concerned solely with digital signals (the more familiar analogue anti-aliasing filter operates on an analogue signal prior to sampling).

We have seen that the decimator consists of an anti-aliasing filter followed by an SRD. In a similar way we can show that an interpolator can be considered as a sampling rate increaser (SRI) followed by a lowpass filter, fig. 5.8(b). The SRI simply inserts $(L-1)$ zero samples between input samples. Its operation and the need for a lowpass filter is deduced from the frequency spectra in fig. 5.10. The input spectrum to the SRI coming

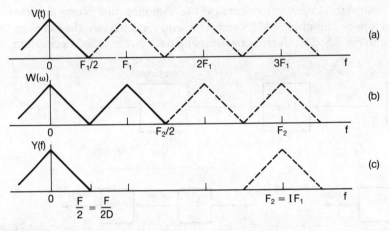

Fig. 5.10   Interpolator spectra

from the decimator with $F_1/2 = B$, fig. 5.10(a), consists of the baseband (or fundamental band) centred at zero frequency, followed by the repeated replicas at $F_1, 2F_1, 3F_1, \ldots$ shown by dashed lines. Assuming that the SRI operates at $F_2 = 3F_1$, its output spectrum $W(\omega)$ is as shown in fig. 5.10(b); the baseband now extends up to $F_2/2$, and the first repetition is centred at F_2 (dashed lines). To obtain the original signal we have to filter out the band between $F_1/2$ and $F_2/2$, resulting in the spectrum of fig. 5.10(c). This filter is called an anti-imaging filter and should be compared to the anti-aliasing filter used in decimation; they are identical for $I = D$.

The decimator and interpolator can be realized with a FIR or IIR filter structure incorporating SRD or SRI units. This section gives only a brief introduction to multirate signal processing. More information can be found in Proakis and Manolakis(11) and Van den Enden(28). Advanced theoretical treatment is given in Crochiere & Rabiner(29).

5.7   Problems

5.1   Assume we have a finite length signal of duration 250 ms, sampled at 512 equally spaced points. Determine the following quantities of its discrete frequency spectrum:
(a) the increment in Hz between successive frequency components;
(b) the repetition period of the spectrum;
(c) the highest frequency permitted in the spectrum of signal to avoid interspectra (or aliasing) interference.

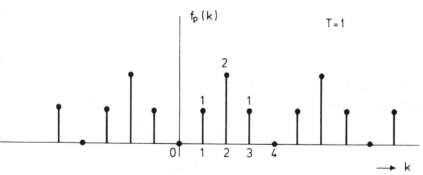

Fig. 5.11

5.2 Calculate the frequency spectrum $F_p(m\Omega)$ for the periodic sequence shown in fig. 5.11, using equation 5.13, and write the z-transform, $F(z)$, for the first period ($k = 0$ to 3) of the sequence in fig. 5.11. Show that at points $\omega = m\Omega$, where Ω is given by equation 5.10, $F_p(m\Omega)$ and $F(z = e^{jm2\pi/N})$ are identical.

5.3 (a) Calculate the magnitude and phase of the frequency response for the periodic time sequence shown in fig. 5.12.
(b) Repeat the calculations of part (a) with the time origin in fig. 5.12 moved to $k = 2$. Compare the results obtained with the ones in part (a).

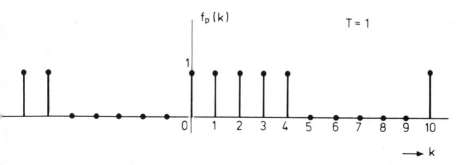

Fig. 5.12

5.4 (a) Consider a discrete signal $f(k)$ whose first three samples are $f(0) = -\frac{1}{2}, f(1) = 1$, $f(2) = -\frac{1}{2}$. Compute its DFT using equation 5.18 and plot the magnitude of the frequency response $|F(m\Omega)|$ for $m = 0, 1, 2, 3$.
(b) Add five zeros to the end of the data given in part (a), i.e. $f(3) = f(4) = \ldots = f(7) = 0$, and calculate the DFT of the new (augmented) sequence. Plot the magnitude $|F(m\Omega)|$ for $m = 0, 1, 2, \ldots, 8$, and show that the frequency resolution has increased. Note: This simple technique allows arbitrary resolution in computing the Fourier transform, as shown by Rabiner & Gold(13), pp. 54–5.

5.5 Given the sequence $g = (2, 1)$ (a) find the two-length and three-length approximate inverse and (b) compute the error squares for each case.

5.6   Compute the three-length approximate inverse for each of the sequences

$$g_1 = (2, 1, 1)$$
$$g_2 = (1, 1, 2)$$

and find the error energy for each case.

5.7   Show that the equations 5.42 can be written in the following generalized manner:

$$r(0)h_0 + r(1)h_1 = p(0)$$
$$r(1)h_0 + r(0)h_1 = p(1)$$

where the coefficients r and p are defined as follows (see, for example, Robinson(8), p. 120): the coefficients r are the autocorrelations of the input sequence to the network h, defined in this case as

$$r(k) = \sum_{i=-\infty}^{\infty} g(i)g(i+k)$$

The coefficients p, are the cross-correlations of the input and desired output sequence of network h, defined in this case as

$$p(k) = \sum_{i=-\infty}^{\infty} g(i)d(i+k)$$

where $d(i)$ is desired output of network h: in our case $d = \delta = (1, 0)$.

Part 2 – Optimum (Wiener and Kalman) linear estimation

Introduction

This part of the book deals with extraction of signal from noisy measured data. The signal usually occupies a limited frequency range while the noise is spread over a wide band of frequencies. In order to remove at least partly the noise from the signal we would use some kind of filtering. In part 2, such filters are derived as estimators of signals in noise. These filters, nonrecursive (FIR) and first-order recursive (IIR) are the structures on which the estimation theory presented here is based. This model-based approach may be more attractive to electrical engineering readers than the alternative based on the linear algebra concepts.

The nonrecursive (FIR) model is used in chapter 6, as a simple-mean estimator for which the mean-square error is established as a measure of the estimation quality. This measure is then taken as the fundamental criterion to derive the optimum FIR filter referred to as the Wiener FIR filter. The result obtained is further extended to derive the Wiener IIR filter. Various features of the solutions are then discussed including an adaptive implementation to determine the FIR filter in an unknown environment.

The recursive (IIR) model is used, in chapter 7, to obtain a suitable type of signal estimator and the mean-square error criterion is again used as the estimation quality measure. This measure is further used, as in the previous chapter, as the fundamental criterion for deriving the optimum IIR type filter. The result is then formulated in a standard algorithmic form known as the scalar Kalman filter.

All the material up to this point refers to the single signal which is termed the scalar (or one-dimensional) signal. In chapter 8, we deal with the vector (or multidimensional) signals, where the results for the scalar Kalman filter are extended to the vector Kalman filter using an equivalence between scalar and matrix operations. We have not extended the Wiener filter from scalar to vector signals, because such an extension is more complicated and not as useful as the vector Kalman filter.

Problems are not given at the end of each chapter in part 2, because the material is such that examples are either too simple or too complicated. However, throughout chapters 6 to 8, a number of examples are discussed in appropriate places. Also, there is a collection of suitable examples with solutions in chapter 9, which illustrate applications of the theory developed in earlier chapters.

6

Nonrecursive (FIR Wiener) estimation

6.0 Introduction

In this chapter we first use the nonrecursive (i.e. FIR) filter to estimate a signal from its noisy samples. These data are often recorded on a magnetic tape or other storage medium before any data processing is done. In such cases the restriction to causal filters is not necessary. However, there are situations in which the data are processed and used as they become available, and in such cases the restriction to causal filters ($h(i) = 0$, for $i < 0$) is normal.

The quality of estimation is assessed in terms of the mean-square error criterion. This is further used to optimize the FIR filter estimator by minimizing the mean-square error. The resulting filter is the FIR Wiener filter, which is then extended to the IIR Wiener filter. Next, a geometric interpretation of the Wiener filter is presented and it is indicated how its bandwidth changes with noise. Finally, we introduce an adaptive implementation of the Wiener FIR filter leading to the well-known least mean-square (LMS) algorithm.

6.1 Nonrecursive (FIR) filtering of noisy data

We use the notation x for a constant input signal, and $x(k)$ for the time varying signal. Measurement of this signal, denoted by $y(k)$, is linearly related to the signal x but it has an additive noise component $v(k)$, introduced by random errors in measurements or any other causes. Therefore, we have

$$y(k) = x + v(k) \tag{6.1}$$

The signal considered here is a random variable with some mean value $E(x) = x_0$ and variance σ_x^2, where $E(x^2) = \sigma_x^2 + [E(x)]^2$ is denoted by S. The noise samples are assumed to be of zero-mean with identical variances σ_x^2, and also to be uncorrelated.

It is assumed that m data samples, specified by equation 6.1, are to be processed by the nonrecursive filter structure of fig. 6.1, with all m weights equal to $1/m$. The input is $y(k)$, and the output is taken as an estimate of parameter x, denoted by $\hat{x}$. Note that the data $y(i)$, $i = 1, 2, \ldots, m$, are available as a *batch*. They are stored, multiplied by equal weights, and the result is summed to produce the output

$$\hat{x} = \frac{1}{m} \sum_{i=0}^{m} y(i) \tag{6.2}$$

In general, the nonrecursive filter processor with different weights is written as

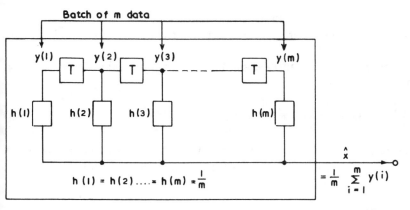

Fig. 6.1   Sample mean estimator

$$\hat{x} = \sum_{i=1}^{m} h(i)\, y(i) \tag{6.3}$$

This is a modified convolution equation (5.27) with inputs $y(i)$ and outputs $\hat{x}(k)$, i.e.

$$\hat{x}(k) = \sum_{i=0}^{m} h(i)\, y(k-i) \tag{6.4}$$

For the finite length filter (m) and observation time $k>m$, one can show graphically that we can leave k out, and equation 6.4 reduces to equation 6.3. In fact, for the equal weights case of equation 6.2 it does not even matter in which order the data within a batch are taken. However, for the case of varying weights, as in equation 6.3, data are processed in an orderly time sequence.

The nonrecursive processor of equation 6.2 is the familiar *sample mean estimator* which we often use as a first approximation to estimate a quantity from m data.

Now, define the error between this estimate and the actual value of x, as $e = \hat{x} - x$. Then, the mean-square error is given by

$$p_e = E(e^2) = E(\hat{x}-x)^2 = E\left[\frac{1}{m}\sum_{i=1}^{m} y(i) - x\right]^2$$

where $y(i) = x + v(i)$ since we assumed additive noise samples $v(i)$.

The above is further written as

$$p_e = E\left\{\frac{1}{m}\sum_{i=1}^{m} [x+v(i)] - x\right\}^2$$

$$= E\left\{\frac{1}{m}\sum_{i=1}^{m} x + \frac{1}{m}\sum_{i=1}^{m} v(i) - x\right\}^2 = E\left[\frac{1}{m}\sum_{i=1}^{m} v(i)\right]^2$$

and finally as

$$p_e = \frac{\sigma_v^2}{m} \tag{6.5}$$

(see appendix 6). This result shows that as the number of samples m increases the mean-square error p_e decreases. Therefore, the sample mean estimate is a good estimate of x in this sense.

The other interesting property of this estimator is obtained by taking the expectation of $\hat{x}$, in equation 6.2, which gives

$$E(\hat{x}) = E\left\{\frac{1}{m}\sum_{i=1}^{m}[x+v(i)]\right\} = E(x) = x_0 \tag{6.6}$$

since, as stated earlier, $E(x) = x_0$, and $E[v(i)] = 0$. The estimate of x, on the average is the same as the average of x. Such an estimator is called an *unbiased estimator*.

6.2   Optimum Wiener FIR filter

We have assumed in the previous section that the mean-square error is a useful criterion of how good an estimation process is. In this section, the mean-square error is taken as the fundamental criterion. The estimates that minimize the mean-square error are taken as the 'best' or optimum estimates. They are also referred to as the least mean-square (LMS) estimates.

We deal with a single signal called a scalar (or one-dimensional) signal. Its estimate is given as the output of a nonrecursive (FIR) filter expressed as

$$\hat{x}(k) = \sum_{i=0}^{m-1} h(i)y(k-i) \tag{6.7}$$

where $y(0), y(1), \ldots, y(m-1)$ are m data signals, and the weights to be determined are $h(i)$. We have taken, in section 6.1, that all weights (or coefficients) are equal. Now we want to choose these weights, $h(i)$, $i = 0, 1, \ldots, m-1$, in such a way that the mean-square error

$$p_e = E(e^2) = E(x-\hat{x})^2$$

is minimized. Note that x is desired signal, and $\hat{x}$ is its estimate, in this case, given by equation 6.7. Substituting for $\hat{x}$, we have

$$p_e = E[x(k) - \sum_{i=0}^{m-1} h(i)y(k-i)]^2 \tag{6.8}$$

The least (or minimum) mean-square error is obtained by differentiation of the above expression with respect to each of m parameters. This is written in the following way:

$$\frac{\partial p_e}{\partial h(j)} = -2E[x(k) - \sum_{i=0}^{m-1} h_0(i)y(k-i)]y(k-j) = 0 \tag{6.9}$$

or $$\sum_{i=0}^{m-1} h_0(i)E[y(k-i)y(k-j)] = E[x(k)y(k-j)] \qquad j = 0, 1, \ldots, m-1 \tag{6.10}$$

The solutions of equation 6.9 are the optimum weights, $h_0(i)$ with $i = 0, 1, \ldots, m-1$.

Before we proceed, we note that equation 6.9 can also be written as

$$E[ey(k-j)] = 0 \qquad j = 0, 1, \ldots, m-1 \tag{6.11}$$

where $e = x - \hat{x}$ is the error. This is referred to as the *orthogonality equation* in estimation theory, to be met again later. It means that the product of the error with each of the measured samples is equal to zero in an expectation (or average) sense.

Returning to equation 6.10, we introduce

$$E[y(k-i)y(k-j)] = p_{yy}(i,j) \tag{6.12}$$

which is the *data autocorrelation function* usually denoted as $R_{yy}(i,j)$, but here we use p_{yy} notation in agreement with generally accepted notation in estimation theory (Kalman filtering particularly). The other point to mention is that $p_{yy}(i,j)$ refers to the non-stationary cases, while for the stationary cases we have $p_{yy}(i-j)$, see, for example, Schwartz & Shaw(27). Similarly we introduce

$$E[x(k)y(k-j)] = p_{xy}(j) \tag{6.13}$$

which is the *cross-correlation* between the random variables $x(k)$ and date $y(k-j)$. Using equations 6.12 and 6.13 in equation 6.10 we have

$$\sum_{i=0}^{m-1} h_0(i) p_{yy}(i-j) = p_{xy}(j) \qquad j = 0, 1, \ldots, m-1 \tag{6.14}$$

Expanding over $i = 0, 1, \ldots, m-1$, this becomes

$$p_{yy}(-j)h_0(0) + p_{yy}(1-j)h_0(1) + \ldots + p_{yy}(m-1-j)h_0(m-1) = p_{xy}(j)$$

and expanding over $j = 0, 1, \ldots, m-1$, we obtain the following set of linear equations:

$$\begin{aligned}
p_{yy}(0)h_0(0) + p_{yy}(1)h_0(1) + \ldots + p_{yy}(m-1)h_0(m-1) &= p_{xy}(0) \\
p_{yy}(-1)h_0(0) + p_{yy}(0)h_0(1) + \ldots + p_{yy}(m-2)h_0(m-1) &= p_{xy}(1) \\
\vdots \qquad\qquad \vdots \qquad\qquad \vdots & \\
p_{yy}(-m+1)h_0(0) + p_{yy}(-m)h_0(1) + \ldots + p_{yy}(0)h_0(m-1) &= p_{xy}(m-1)
\end{aligned} \tag{6.15}$$

where $p_{yy}(k) = p_{yy}(-k)$ since p_{yy} is symmetrical. The *known* quantities are $p_{yy}(i,j)$, i.e. the autocorrelation coefficients of the input data, and $p_{xy}(j)$, the cross-correlation coefficients between desired signal x and the data y. The *unknown* quantities $h_0(i)$, $i = 0, 1, \ldots, m-1$, are the optimum filter coefficients.

The least mean-square error corresponding to the above optimum solution is obtained from

$$p_{e0} = E(e^2) = E\left\{ e[x(k) - \sum_{i=0}^{m-1} h_0(i)y(k-i)] \right\} = E\{ex(k)\}$$

where equation 6.11 has been applied. Therefore, we have

$$p_{e0} = E(x^2) - \sum_{i=0}^{m-1} h_0(i) E[x(k)y(k-i)]$$

or

$$p_{e0} = p_{xx}(0) - \sum_{i=0}^{m-1} h_0(i) p_{xy}(i) \tag{6.16}$$

where we have used equation 6.13, and also introduced the signal autocorrelation for zero lag, $p_{xx}(0) = E(x^2)$.

The complete solution to the estimation problem, in this case, is given by the set of equations 6.15, estimator equation 6.7, and the corresponding least mean-square error, equation 6.16.

Matrix forms of these three equations are

$$P_{yy} \mathbf{h}_0 = \mathbf{p}_{xy} \tag{6.15'}$$

where P_{yy} is $m \times m$ correlation matrix, $\mathbf{h}$ and $\mathbf{p}_{xy}$ are $m \times 1$ column vectors. The formal solution of equation 6.15′ is

$$\mathbf{h}_0 = P_{yy}^{-1} \mathbf{p}_{xy} \tag{6.17}$$

On the other hand the estiimate can be written as

$$\hat{x}(k) = \mathbf{h}_0^{\mathrm{T}} \mathbf{y}(k) = p_{xy}^{\mathrm{T}} P_{yy}^{-1} \mathbf{y}(k) \tag{6.7'}$$

and the least mean-square error

$$p_{e0} = p_{xx}(0) - \mathbf{p}_{xy}^{\mathrm{T}} P_{yy}^{-1} \mathbf{p}_{xv} \tag{6.16'}$$

since the matrix P_{yy} is symmetrical.

A filter of the type described above is often called a scalar (or one-dimensional) Wiener filter, and equation 6.14 is known as the *Wiener-Hopf equation.*

It should be noted that the measurement equation $y(k) = x(k) + v(k)$ has not been used in the above equations. Therefore, the result is more general than it appears. Hence, if the data samples $y(i)$, $i = 0, 1, \ldots, m - 1$ 'somehow' contain the unknown random variable $x(k)$, the signal, the best *linear* filter operation on the samples to estimate $x(k)$ is given by the Wiener filter.

Example 6.1
A signal x has been measured m times in the presence of additive noise, i.e. we have $y(i) = x + v(i), i = 0, 1, \ldots, m - 1$. It is assumed that the noise samples are of zero mean and variance σ_v^2, uncorrelated with each other and with the signal x as well. This is expressed by

$$E[v(j)v(k)] = \begin{cases} \sigma_v^2, & j = k \\ 0, & j \neq k \end{cases}$$

and $E[xv(j)] = 0$. Furthermore, we assume $E(x) = 0$, and hence $E(x^2) = \sigma_x^2$.

To solve this, we have to calculate: $p_{yy}(i, j)$ and $p_{xy}(j)$.

$$p_{yy}(i, j) = E[y(i)y(j)] = E\{[x + v(i)][x + v(j)]\} = \sigma_x^2 + \sigma_v^2 \delta(i, j)$$

where $\delta(i, j) = 1$, for $i = j$, $\delta(i, j) = 0$ for $i \neq j$.

$$p_{xy}(j) = E[xy(j)] = E(x^2) = \sigma_x^2$$

Substituting these quantities into equations 6.15 and solving we obtain

$$h_0(0) = h_0(1) = \ldots = h_0(m - 1) = 1/(m + \gamma)$$

where $\gamma = \sigma_v^2 / \sigma_x^2$ is the noise-to-signal ratio.

The estimator, equation 6.7, for this case becomes

$$\hat{x} = \frac{1}{m+\gamma} \sum_{i=0}^{m-1} y(i)$$

and the corresponding error

$$P_e = \frac{\sigma_v^2}{m+\gamma}$$

Note that for large signal-to-noise ratio we have $\gamma \ll m$, and the above estimator reduces to the sample mean estimate discussed in section 6.1.

Example 6.2

Assume two samples of a linearly increasing signal have been measured in the presense of independent additive noise. Estimate the slope x of this straight line by means of the optimum linear processor

$$\hat{x} = \sum_{i=1}^{2} h(i)y(i)$$

Assume noise samples to be uncorrelated, with variance σ_v^2.

From the above we can write the data equation as

$$y(k) = kx + v(k) \qquad k = 1, 2$$

where kx is a straight line, and $v(k)$ is additive noise. To solve the problem we have to determine $p_{yy}(i,j)$ and $p_{xy}(j)$ for equations (6.15):

$$\begin{aligned}
p_{yy}(i,j) &= E[y(i)y(j)] = E\{[ix+v(i)][jx+v(j)]\} \\
&= ijE(x^2) + E[v(v(j)] = ijS + \sigma_v^2 \delta(i,j) \qquad i,j = 1, 2 \\
S &= E(x^2) = \sigma_x^2 + [E(x)]^2 \\
p_{xy}(j) &= E[xy(j)] = E\{x[jx+v(j)]\} = jS \qquad j = 1, 2
\end{aligned}$$

Substituting these quantities into equation 6.15 and solving for h's we obtain:

$$h_0(1) = \frac{1}{5+\gamma}, \; h_0(2) = \frac{2}{5+\gamma}$$

where $\gamma = \sigma_v^2/S$. The estimate is then, from equation 6.7, given by

$$\hat{x} = \frac{y(1) + 2y(2)}{5+\gamma}$$

and the least mean-square error, from equation 6.16, is

$$P_{eo} = S \frac{\gamma}{5+\gamma}$$

Example 6.3

Consider the signal $y(k) = x(k) + v(k)$ consisting of a desired signal $x(k)$ with the autocorrelation sequence $p_{xx}(n) = 0.8^{|n|}$, $n = 0, \pm 1, \pm 2, \ldots$ and a zero-mean white noise $v(k)$ with variance 0.64. The signal $x(k)$ and the noise $v(k)$ are statistically independent.

Design a FIR filter of length 3 to process $y(k)$ so that its output $\hat{x}(k)$ minimizes $E[\hat{x}(k) - x(k)]^2$

For this problem we have $p_{yy}(n) = p_{xx}(n) + p_{vv}(n) = 0.8^{|n|}$, for $n \neq 0$, $p_{yy}(0) = 1 + 0.64 = 1.64$, and $p_{xy}(n) = p_{xx}(n)$.

Hence, equation 6.15 becomes

$$\begin{bmatrix} 1.64 & 0.8 & 0.64 \\ 0.8 & 1.64 & 0.8 \\ 0.64 & 0.8 & 1.64 \end{bmatrix} \begin{bmatrix} h_0(0) \\ h_0(1) \\ h_0(2) \end{bmatrix} = \begin{bmatrix} 1 \\ 0.8 \\ 0.64 \end{bmatrix}$$

Solutions of this set of equations are:

$$h_0(0) = 0.421, \, h_0(1) = 0.188, \, h_0(2) = 0.096$$

Hence, the optimum filter transfer function is

$$H_0(z) = 0.421 + 0.188z^{-1} + 0.096z^{-2}$$

The mean-square error of this filter is

$$p_{eo} = p_{xx}(0) - \sum_{i=0}^{2} h_0(i)p_{xy}(i)$$

where, in this case, $p_{xy}(i) = p_{xx}(i)$.

$$\therefore \quad p_{e0} = 1 - (0.421 \times 1 + 0.188 \times 0.8 \times 0.096 \times 0.64) = 0.367$$

For the same case, but with $\sigma_v^2 = 0.49$, Wiener filter weights are

$$h_0(0) = 0.519, \, h_0(1) = 0.21, \, h_0(2) = 0.095, \text{ and } p_{e0} = 0.2522.$$

6.3 IIR Wiener filter

The optimum IIR Wiener filter is obtained from equation 6.14 by setting the upper limit $m = \infty$

$$\sum_{i=0}^{\infty} h_0(i)p_{yy}(i-j) = p_{xy}(j) \tag{6.18}$$

The impulse response is now of infinite duration, hence it cannot be obtained in the time domain. However, solution can be obtained by transformation of equation 6.18 into z-domain. This cannot be done directly because the impulse response for causal (realizable) systems is zero for negative time. But, we can proceed by considering non-causal systems for which equation 6.18 becomes

$$\sum_{i=-\infty}^{\infty} h_{ou}(i)p_{yy}(i-j) = p_{xy}(j) \tag{6.19}$$

where h_{ou} is the impulse response of the optimum unrealizable filter. The left-hand side of equation 6.19, being convolution of two sequences $h_{ou}(i)$ and $p_{yy}(i)$, transforms into multiplication of their respective z-transform:

$$H_{ou}(z)S_{yy}(z) = S_{xy}(z) \tag{6.20}$$

where

$$S_{yy}(z) = \sum_{k=-\infty}^{\infty} p_{yy}(k)z^{-k} \tag{6.21}$$

is *the power spectral density* $S_{yy}(\omega)$ when we set $z = e^{j\omega}$, and similarly $S_{xy}(z)$ corresponds to the cross-spectral density, and

$$H_{ou}(z) = \sum_{i=-\infty}^{\infty} h_{ou}(i)z^{-i} \tag{6.22}$$

From equation 6.20, the transfer function of the IIR Wiener filter is

$$H_{ou}(z) = \frac{S_{xy}(z)}{S_{yy}(z)} \tag{6.23}$$

Such a non-causal filter is physically unrealizable for real-time operation. This solution is of interest for extension to realizable filters, and it is also of interest when non-real-time operation is admissible.

Before we proceed, it is useful to indicate some properties of the function $S_{yy}(z)$. It can be written as

$$S_{yy}(z) = \sigma^2 S_y^+(y) S_y^-(z) \tag{6.24}$$

where σ^2 denotes a constant obtained by power spectrum factorization, the function $S_y^+(z)$ has no poles or zeros for $|z| > 1$ i.e. outside the unit circle, and the function $S_y^-(z)$ has no poles or zeros for $|z| < 1$ i.e. within the unit circle. This is illustrated in the following example

Example 6.4
Suppose $p_{yy}(n) = a^{|n|}$ with $0 < a < 1$. Then,

$$S_{yy}(z) = \sum_{n=-\infty}^{\infty} a^{|n|}z^{-n} = \sum_{n=-\infty}^{0} a^{-n}z^{-n} + \sum_{n=0}^{\infty} a^n z^{-n} - 1 \tag{6.25}$$

The first summation, over negative time $n < 0$, results in the function $1/(1 - az)$ with the pole at $z = 1/a$ (>1). The second summation, over positive time $n > 0$, results in the function $1/(1 - az^{-1})$, with the pole at $z = a$ (<1).

Combining these two parts we have

$$s_{yy}(z) = \frac{1 - a^2}{(1 - az^{-1})(1 - az)} \tag{6.26}$$

$$\therefore \quad S_y^+(z) = \frac{(1 - a^2)^{1/2}}{1 - az^{-1}} \quad \text{(Pole inside unit circle; valid for } n > 0)$$

$$S_y^-(z) = \frac{(1 - a^2)^{1/2}}{1 - az} \quad \text{(Pole outside unit circle; valid for } n < 0)$$

We can now proceed to derive a realizable IIR Wiener filter by using equation 6.24 in equation 6.23 which gives

$$H_{ou}(z) = \frac{S_{xy}(z)}{\sigma^2 S_y^+(z) S_y^-(z)}$$

The causal (realizable) IIR optimum Wiener filter can now be stated as

$$H_o(z) = \frac{1}{\sigma^2 S_y^+(z)} \left[\frac{S_{xy}(z)}{S_y^-(z)} \right]_{cp} \tag{6.27}$$

where the 'cp' notation signifies the 'causal part' of $[.]$. This represents the z-transform version of the solution to the Wiener-Hopf equation (6.18).

The minimum mean-square error is obtained from the FIR result by setting the upper limit in equation 6.16 to be ∞, i.e.

$$p_{e0} = p_{xx}(0) - \sum_{i=0}^{\infty} h_0(i) p_{xy}(i) \tag{6.28}$$

For a filter whose $H_0(i)$ is obtained from equation 6.27, one can apply the inverse z-transform to obtain $h_0(k)$, and use it in equation 6.28.

Example 6.5
As in example 6.3, the measured signal is $y(k) = x(k) + v(k)$, where the desired signal has the autocorrelation function $p_{xx}(n) = 0.8^{|n|}$, and the noise has zero mean and variance $\sigma_v^2 = 0.64$. They are statistically independent. It is required to find the optimum IIR Wiener filter.

Using the result of example 6.4, with $a = 0.8$, we have immediately

$$S_{xx}(z) = \frac{0.36}{(1-0.8z^{-1})(1-0.8z)} = \frac{-0.45z}{(z-0.8)(z-1.25)}$$

and $S_{vv}(z) = 0.64$

$\therefore$ $S_{yy}(z) = S_{xx}(z) + S_{vv}(z) = 0.64 \dfrac{z^2 - 2.753z + 1}{(z-0.8)(z-1.25)}$

and $S_{xy} = S_{xx}(z)$

Zeros of $S_{yy}(z)$ are $z_1 = 2.322$ and $z_2 = 0.431$. Therefore

$$S_{yy}(z) = \underbrace{0.64}_{\sigma^2} \; \underbrace{\left(\frac{z-0.431}{z-0.8}\right)}_{S_y^+(z)} \underbrace{\left(\frac{z-2.322}{z-1.25}\right)}_{S_y^-(z)}$$

$$\frac{S_{xy}(z)}{S_y^-(z)} = \frac{-0.45z}{(z-0.8)(z-1.25)} \frac{z-1.25}{z-2.322}$$

Performing the partial fraction expansion we have

$$\frac{S_{xy}(z)}{S_y^-(z)} = A\frac{z}{z-0.8} + B\frac{z}{z-2.322}$$

where $A = 0.2957$ and B is not important since its pole is outside the unit circle. Therefore

$$\left[\frac{S_{xy}(z)}{S_y^-(z)}\right]_{cp} = 0.2957\frac{z}{z-0.8}$$

Using this result and S_y^+ from $S_{yy}(z)$ in equation 6.27, we obtain

$$H_0(z) = \frac{0.462}{1-0.431z^{-1}}$$

or in the time domain: $h(n) = 0.462(0.431)^n$.

The first three values are $h(0) = 0.462$, $h(1) = 0.199$, $h(2) = 0.086$. These compare well with the values obtained in example 6.3 for the Wiener FIR filter.

The minimum mean-square error is given by

$$p_{e0} = p_{xx}(0) - \sum_{k=0}^{\infty} h(k)p_{xy}(k) = 1 - \sum_{k=0}^{\infty} 0.462(0.431)^k (0.8)^k$$

$\therefore$ $\quad \underline{p_{e0} = 0.295}$

Comparing p_{e0} obtained here with p_{e0} in example 6.3, we see that p_{e0} has improved. Also, the filter structure is simpler since $H_0(z)$ realization requires only one delay unit. For $\sigma_v^2 = 0.49$, $H_0(z) = 0.516/(1-0.387z^{-1})$ hence $h(n) = 0.516(0.387)^n$, and $p_{e0} = 0.2523$. (Here p_{e0} has not improved in comparison with the same case in example 6.3.)

6.4 Comments on Wiener filters

It is interesting to note that in example 6.5, we have a first-order signal described by

$$S_y^+(z) = \frac{A}{a - az^{-1}} \qquad 0 \le a \le 1 \tag{6.29}$$

The transfer function of the Wiener IIR filter has been found to be

$$H(z) = \frac{B}{1 - bz^{-1}} \qquad 0 \le b \le 1 \tag{6.30}$$

This means that the filter has a spectral shape similar to the signal. Its frequency response is given by

$$H(\omega) = H(z = e^{j\omega T}) = \frac{B e^{j\omega T}}{e^{j\omega T} - b}$$

or $\quad \left| \frac{H(\omega)}{H(0)} \right| = (1 - b)\frac{|e^{j\omega T}|}{|e^{j\omega T} - b|}$ \hfill (6.31)

Therefore, the magnitude of the frequency response, at frequency ω, is the ratio of the length of the vector $v_1 = e^{j\omega T}$ and the vector $v_2 = e^{j\omega T} - b$, as shown in fig. 6.2(a), i.e. $|H(\omega)|H(0)| \propto OL/NL$, where $OL = 1$, and NL is the pole vector (scaling factor $(1 - b)$ is ignored in this discussion).

At $\omega = 0$, (NL_0) is minimum, and then increases with ω. The frequency at which $\overline{NL_1} = 2(\overline{NL_0})$ is the $(-3\,\text{dB})$ frequency. If b is close to 1, then the $(-3\,\text{dB})$ cutoff frequency is low, i.e. the system is narrowband, fig. 6.2(b). When b decreases, then the system changes to a wideband type, fig. 6.2(c). There is an indication of this behaviour in example 6.5, where for $\sigma_v^2 = 0.64$, $b = 0.431$, and for $\sigma_v^2 = 0.49$, $b = 0.387$. In a more general analysis of the first order system with additive white noise, one can show (Bozic(30)) that the value of b depends on both the signal parameter a, and the signal-to-noise ratio (R). Then, for a fixed value of a, and increased R one finds that b decreases, i.e. the system passband widens to allow more signal to pass through.

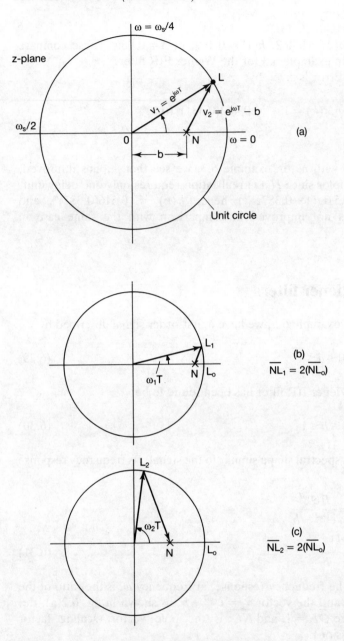

Fig. 6.2   IIR Wiener filter geometric interpretation

However, when R decreases, the value of b increases, making the system narrowband, trying to eliminate as much noise as possible. In fact, it appears that the Wiener IIR filter, for this case of a lowpass type signal, is also a lowpass filter whose cutoff frequency is a function of the signal-to-noise ratio. Similar behaviour is found for the second-order signals.

From the computational point of view, the Wiener FIR filter requires solving of a set of linear equations. For a larger number of points it is then convenient to use matrices, and also recursive methods for solution of the Wiener-Hopf equation.

The main problem to solve in the Wiener IIR filters is the spectral factorization illustrated in example 6.5. This can become in general quite involved, particularly in a changing environment, so that the approximation of an IIR filter by a finite length FIR filter is preferable.

However, there are difficulties with the Wiener FIR filter listed below:

(i)   it requires knowledge of the autocorrelation sequence of the filter input, and the cross-correlation sequence between the filter input and some desired response;

(ii)   the number of data samples m, must be specified, and if m is charged (more data available), the calculations must be repeated;

(iii)   when the filter operates in an unknown environment the information in the above (i) and (ii) is not available.

Therefore, we have to find an adaptive implementation of the filter. The term adaptive means to be able to learn from its environment and hence adjust the coefficients in a recursive manner towards the optimum values. An adaptive implementation of the Wiener filter is discussed in the next section. An alternative is to use the recursive estimator or Kalman filter which is introduced in the following chapter.

6.5   Adaptive implementation of the Wiener filter

The first step is to formulate a recursive solution for the Wiener-Hopf (or normal) equations (6.14). This is done by varying the FIR filter coefficients with time in a *recursive* manner as follows

$$h(k, n+1) = h(k, n) - \frac{1}{2} \mu \nabla_k(n) \tag{6.32}$$

where k refers to k-th filter coefficient, n is discrete time iteration, μ is a constant, and

$$\nabla_k(n) = \frac{\partial p_e(n)}{\partial h(k, n)} \tag{6.33}$$

is the *gradient of the mean-square error (or cost) function* with respect to k-th filter coefficient. Since $p_e(n) = E[e^2(n)]$, we have

$$\frac{\partial p_e}{\partial h(k, n)} = 2E\left[e(n)\frac{\partial e(n)}{\partial h(k)}\right] \tag{6.34}$$

where

$$e(n) = d(n) - \sum_{k=0}^{m-1} h(k)y(n-k)$$

$$\frac{\partial p_e(n)}{\partial h(k, n)} = -2E[e(n)y(n-k)]$$

$$\therefore \qquad \nabla_k(n) = -2p_{ey}(k) \tag{6.35}$$

Substituting equation 6.35 into equation 6.32 we obtain

$$h(k, n+1) = h(k, n) + \mu p_{ey}(k) \tag{6.36}$$

This result means that we may compute the *updated* value of the k-th filter coefficient by applying a *correction* to the previous value $h(k, n)$ of this coefficient.

The correction term in equation 6.36 should approach zero as the number of iterations n approaches infinity, so that *in the limit $h(k, n)$ approaches the optimum Wiener filter coefficient $h_0(k)$*. The convergence of p_{ey} towards zero corresponds to the orthogonality equation 6.11 being zero.

It can be shown (Candy(31)) that the expectation (or averaging) operation can be avoided by using an *instantaneous estimate* for the cross-correlation $p_{ey}(z)$:

$$\hat{p}_{ey}(z) = e(n)y(n-k)$$

This estimate is unbiased in the sense that its mean equals the actual value of $p_{ey}(k)$. Denoting by $\hat{h}(k, n)$ the corresponding estimate of the filter coefficient $h(k, n)$, we can rewrite equation 6.36 as

$$\hat{h}(k, n+1) = \hat{h}(k, n) + \mu e(n)y(n-k)$$

We can now write *the least mean-square (LMS) algorithm* as follows:

$$\hat{x}(n) = \sum_{k=0}^{m-1} h(k, n)y(n-k) \tag{6.37}$$

$$e(n) = x(n) - \hat{x}(n) \tag{6.38}$$

$$\hat{h}(k, n+1) = \hat{h}(k, n) + \mu e(n)y(n-k) \tag{6.39}$$

where $x(n)$ is the desired response, and $\hat{x}(n)$ is the filter output at time n.

Typically, the LMS algorithm is *initialized* by setting all coefficients ($k = 0, 1, \ldots, m-1$) at time $n = 0$ to zero. Then, the algorithm proceeds by first computing the error signal $e(n)$, which is used to compute the updated coefficient estimate $h(k, 1)$. The cycle is repeated for $n = 2, 3, \ldots$ until steady-state values are obtained.

The block diagram in fig. 6.3 represents the set of equations 6.37 to 6.39. It shows that an adaptive filter using the LMS algorithm is a *closed-loop system* with time-varying

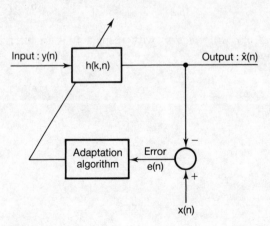

Fig. 6.3   Adaptive implementation of the Wiener filter

parameters. As such, the algorithm can become unstable. The stability of the algorithm depends on the adaptation parameter μ, which has to be chosen carefully. For convergence two requirements have to be satisfied. *First* the filter coefficients approach the Wiener optimum solution p_{e0} in the *mean* as $n \to \infty$, i.e. the number of iterations approaches infinity. *Second*, the average mean-square error (averaging over both the filter input and the filter coefficients) approaches some constant value as $n \to \infty$. In general, this constant value is in excess of the minimum mean-square error computed from the use of the normal equations.

Both of these requirements are satisfied if μ satisfies the following condition:

$$0 < \mu < \frac{2}{\text{total input power}} \tag{6.40}$$

where the 'total input power' refers to the sum of the mean-square values of the tap inputs $y(n)$, $y(n-1)$, $\ldots$, $y(n-m+1)$. For a stationary input, the total input power equals $mp_{yy}(0)$, where m is the number of taps and $p_{yy}(0)$ is the autocorrelation function of the tap inputs for zero lag. Note that the adaptation parameter μ has a dimension that is the inverse of power, which makes the correction term in equation 6.39 dimensionless.

When μ *is small*, the LMS algorithm has more time to learn about its input with the result that the average mean-square error is only slightly in excess of p_{e0}. This is obtained at the cost of a large number of iterations of the algorithm. When μ *is large*, the LMS algorithm reaches its steady-state condition faster, but at the expense of an increase in the average mean-square error.

Example 6.6
The aim is to show how equation 6.32 arises from the work done in section 6.2. We start with equations 6.9, not set to zero, and obtain immediately the gradient of p_e:

$$\nabla_j = \frac{\partial p_e}{\partial h(j)} = -2p_{xy}(j) + 2 \sum_{i=0}^{m-1} h(i) p_{yy}(i, j)$$

where $p_{yy}(i, j)$ and $p_{xy}(j)$ are as defined in equations 6.12 and 6.13 respectively. Expanding the above equation over $j = 0, 1, \ldots, m-1$, we have the vector equation

$$\nabla = 2(P_{yy} \mathbf{h} - \mathbf{p}_{xy})$$

where ∇, $\mathbf{h}$, $\mathbf{p}_{xy}$ are $(m \times 1)$ column vectors and P_{yy} is $(m \times m)$ matrix as used in section 6.2. We rearrange this as

$$\nabla = 2P_{yy}(\mathbf{h} - P_{yy}^{-1} \mathbf{p}_{xy})$$

where the second term within the bracket is $\mathbf{h}_0$ as given in equation 6.17. Therefore, from the above, we can establish the following equation

$$\mathbf{h}_0 = \mathbf{h} - \frac{1}{2} P_{yy}^{-1} \nabla \tag{6.41}$$

This shows that given any weight vector, $\mathbf{h}$, P_{yy} and ∇, we can obtain $\mathbf{h}_0$ from any $\mathbf{h}$ in a single step, by adjusting $\mathbf{h}$ according to equation 6.41. However, in practice the P_{yy} matrix is unknown, and can at best only be estimated. Also, the ∇ vector would have to be estimated at each computing iteration.

Therefore, in practice equation 6.41 is modified into an algorithm 6.32 which adjusts $\mathbf{h}$

in small increments and converges to $\mathbf{h}_0$ after many iterations as briefly introduced in this section.

Numerical examples are not given in this section, because even relatively simple cases require a good deal of computation. Good simple examples can be found in Candy(31) and a more advanced study in Lim & Oppenheim(32).

7
Recursive (Kalman) estimation

7.0 Introduction

Here we use the recursive (IIR) filter for noisy data filtering. As in the previous chapter, we assess how well this filtering operation is performed in terms of the mean-square error. However, now the signal is specified with some precision as a first-order autoregressive process. This is subsequently used to derive the optimum recursive filter or perhaps better known as the scalar Kalman filter. The same procedure is further used to obtain the one-step Kalman predictor, and examine its relationship with the Kalman filter. In the final section, similarly as with the Wiener filter in the previous chapter, the Kalman filter functioning is given a geometric interpretation in the z-domain.

7.1 Recursive (IIR) filtering of noisy data

Consider now the simple first order recursive filter, fig. 7.1, where $y(k)$ and $g(k)$ are the input and output sequences respectively. As in section 6.1, the input signal

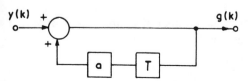

Fig. 7.1 Recursive filter as noisy data processor

$y(k) = x + v(i)$ represents measurements of x in the presence of the additive noise. The filter output is given by

$$g(k) = y(k) + ag(k-1) \tag{7.1}$$

For a sequence of input samples $y(1), y(2), \ldots, y(m)$, and assuming $g(k) = 0$ for $k < 1$, we obtain the following outputs:

$$g(1) = y(1) + ag(0) = y(1)$$
$$g(2) = y(2) + ag(1) = y(2) + ay(1)$$
$$g(3) = y(3) + ag(2) = y(3) + ay(2) + a^2 y(1)$$
$$\vdots$$
$$g(m) = y(m) + ag(m-1) = y(m) + ay(m-1) + a^2 y(m-2) +$$
$$\ldots + a^{m-2} y(2) + a^{m-1} y(1)$$

Substituting for $y(m) = x + v(m)$, and separating the signal and noise terms, we obtain

$$g(m) = (1 + a + a^2 + \ldots + a^{m-1})x + [v(m) + av(m-1) + \ldots + a^{m-1}v(1)]$$

or $\quad g(m) = \dfrac{1 - a^m}{1 - a} x + \sum_{i=1}^{m} a^{m-i} v(i)$ $\qquad\qquad$ (7.2)

where the first term is the sum of the geometric series associated with x. For large m, $|a|^m \ll 1$, and the signal part of $g(m)$ approaches $x/(1-a)$. This indicates that a good estimate of x is given by

$$\hat{x} = (1-a)g(m)$$

i.e. $\quad \hat{x} = (1 - a^m)x + (1-a)\sum_{i=1}^{m} a^{(m-i)} v(i)$ $\qquad\qquad$ (7.3)

This means the output $(1-a)g(m)$ is taken as an estimate of signal x, after the m-th input sample has been processed, fig. 7.2.

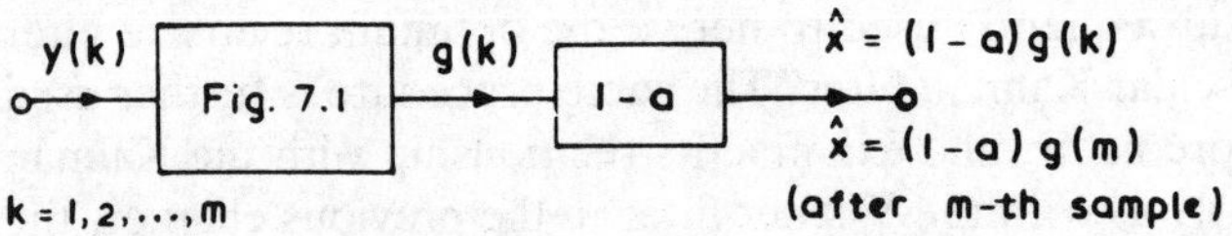

Fig. 7.2 Recursive filter as an estimator

 To assess how good is the *recursive estimator*, we calculate the mean-square error as shown in appendix 7. It is found there that the output mean-square error, after the m-th sample has been processed, is given by

$$p_e = a^{2m} S + \frac{(1-a)(1-a^{2m})}{1+a} \sigma_v^2$$ $\qquad\qquad$ (7.4)

Since $a < 1$, for stability, and by increasing m (i.e. the number of samples) we can obtain $a^{2m} \to 0$. Then, the above estimation error reduces to

$$p_e = \frac{1-a}{1+a} \sigma_v^2 = \sigma_v^2 \Big/ \left(\frac{1+a}{1-a}\right)$$ $\qquad\qquad$ (7.5)

for $a = 0.9$, $(1+a)/(1-a) \cong 20$. The larger a is (but <1), the smaller p_e will be.

7.2 Signal and observation models

We use again the approach developed in chapter 6 i.e. we deal with the one-dimensional (or scalar) signals and use the least mean-square criterion. There, the signal has been described by means of its *autocorrelation sequence*. However, in this section we start with a more precise description of the signal in terms of a *first-order recursive model*.

 We model the signal by a first-order recursive filter driven by zero-mean white noise. Therefore, the signal evolves in time according to the dynamics described by equation

$$x(k) = ax(k-1) + w(k-1)$$ $\qquad\qquad$ (7.6)

also shown in fig. 7.3(a). The random drive is specified by $E[w(k)] = 0$, and

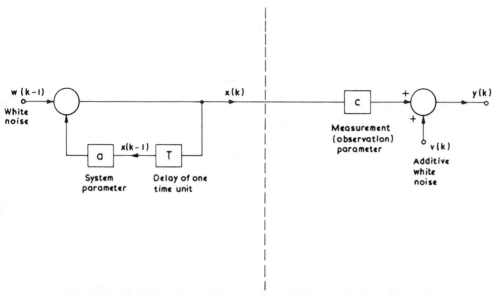

Fig. 7.3   (a) Model of random signal process; (b) measurement (observation) model

$$E[w(k)w(j)] = \begin{cases} \sigma_w^2, & k = j \\ 0, & k \neq j \end{cases} \qquad (7.7)$$

which represents the white noise.

If $\sigma_w^2 = 0$, the noise process disappears, and we obtain a family of deterministic signals shown in fig. 7.4. Adding to this the noise, one obtains a family of random processes.

A random process defined by equation 7.6 is said to be an autoregressive process of the first order. It can be shown that the statistical parameters of $x(k)$ are:

$$\left. \begin{aligned} & E[x(k)] = 0 \\ & E[x^2(k)] = p_{xx}(0) = \sigma_x^2 = \frac{\sigma_w^2}{1 - a^2} \\ & E[x(k)x(k+j)] = p_{xx}(j) = a^{|j|} p_{xx}(0) \end{aligned} \right\} \qquad (7.8)$$

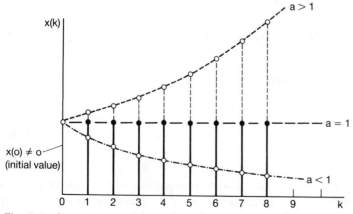

Fig. 7.4   Signal model $x(k) = ax(k-1)$ and responses for various values of a

where j represents the spacing between samples, and $p_{xx}(j)$ is the autocorrelation function. Note that the driving white noise is uncorrelated, but the process generated by this model is correlated as shown by $p_{xx}(j)$. For more detail see, for example, Schwartz & Shaw(27).

The observation model is assumed again to be linear, fig. 7.3(b), described by

$$y(k) = cx(k) + v(k) \tag{7.9}$$

The time-varying random signal is described by equation 7.6, and the factor c represents an observation (or measurement) parameter. It will be seen later that this factor is useful in transforming results to vector signals. As before, $v(k)$ is an independent additive white noise with zero-mean and variance σ_v^2, i.e.

$$E[v(k)] = 0$$

$$E[v(k)v(j)] = \begin{cases} \sigma_v^2, & k = j \\ 0, & k \neq j \end{cases} \tag{7.10}$$

7.3   Optimum recursive estimator (scalar Kalman filter)

The recursive estimator is of the form

$$\hat{x}(k) = a(k)\hat{x}(k-1) + b(k)y(k) \tag{7.11}$$

where the first term represents the weighted previous estimate, and the second term is weighted present data sample. In this case we have two parameters, $a(k)$ and $b(k)$, to be determined from minimization of the mean-square error

$$p(k) = E[e^2(k)]$$

where $e(k) = \hat{x}(k) - x(k)$ is the error. (The same procedure has been used in section 6.2, but the processor there was of nonrecursive type, and the number of coefficients $h(i)$ was equal to the number, m, of data available.)

Substituting equation 7.11 for $\hat{x}(k)$, we have

$$p(k) = E[a(k)\hat{x}(k-1) + b(k)y(k) - x(k)]^2 \tag{7.12}$$

Differentiating with respect to $a(k)$ and $b(k)$ we obtain

$$\frac{\partial p(k)}{\partial a(k)} = 2E[a(k)\hat{x}(k-1) + b(k)y(k) - x(k)]\hat{x}(k-1) = 0 \tag{7.13}$$

$$\frac{\partial p(k)}{\partial b(k)} = 2E[a(k)\hat{x}(k-1) + b(k)y(k) - x(k)]y(k) = 0 \tag{7.14}$$

Alternatively

$$E[e(k)\hat{x}(k-1)] = 0 \tag{7.15}$$

and $$E[e(k)y(k)] = 0 \tag{7.16}$$

which are the *orthogonality equations* corresponding to equation 6.11 in section 6.2.

The first equation of the above set of equations is used to determine the relationship (see, appendix 8) between $a(k)$ and $b(k)$

$$a(k) = a[1 - cb(k)]$$

Applying this to equation 7.11, we have

$$\hat{x}(k) = a\hat{x}(k-1) + b(k)[y(k) - ac\hat{x}(k-1)] \tag{7.17}$$

The first term, $a\hat{x}(k-1)$, represents the best estimate of $\hat{x}(k)$ without any additional information, and it is therefore *a prediction* based on past observations. The second term is *a correction* term involving the difference between the new data sample and the

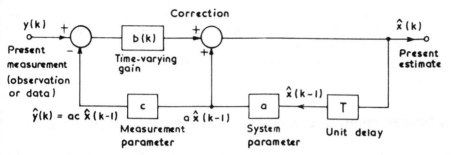

Fig. 7.5 Optimum recursive estimator (filter)

observation estimate, $\hat{y}(k) = ac\hat{x}(k-1)$, weighted by a variable gain factor $b(k)$, fig. 7.5. It is shown in appendix 9 that

$$b(k) = \frac{c[a^2 p(k-1) + \sigma_w^2]}{\sigma_v^2 + c^2 \sigma_w^2 + c^2 a^2 p(k-1)} \tag{7.18}$$

and also that the mean-square error is given by

$$p(k) = \frac{1}{c}\sigma_v^2 b(k) \tag{7.19}$$

i.e. $p(k)$ and $b(k)$ are directly related

Equations 7.17 to 7.19 constitute a complete computational algorithm, and it can be used in this form. However, for the purpose of extending these results later to vector signals, we rearrange these equations as shown below.

Recursive filter estimator:
$$\hat{x}(k) = a\hat{x}(k-1) + b(k)[y(k) - ac\hat{x}(k-1)] \tag{7.20}$$

Filter gain:
$$b(k) = cp_1(k)[c^2 p_1(k) + \sigma_v^2]^{-1} \tag{7.21}$$

where

$$p_1(k) = a^2 p(k-1) + \sigma_w^2 \tag{7.22}$$

Mean-square error:
$$p(k) = p_1(k) - cb(k)p_1(k) \tag{7.23}$$

The estimator equation 7.20 is the same as the original equation 7.17, but equations 7.18 and 7.19 are now written as three equations, 7.21 to 7.23, because we have introduced a new quantity $p_1(k)$. This quantity has an important role, as will be seen later in the next chapter.

The above set of equations constitute *the scalar (or one-dimensional) Kalman filter*, for the signal model, equation 7.6,

$$x(k) = ax(k-1) + w(k-1)$$

and the measurement model, equation 7.9,

$$y(k) = cx(k) + v(k)$$

This type of recursive estimation technique was developed around 1960, most notably by R.E. Kalman(33). For this reason, the processors devised at that time (as well as the wide variety of generalizations and extensions) are referred to as *Kalman filters*. However, there were also other workers in this field who claimed priority, Sorenson(34).

7.4　Optimum recursive predictor (scalar Kalman predictor)

The filtering problem discussed so far means the estimation of the current value of a random signal in the presence of additive noise. It is often required, particularly in control systems, to predict ahead it possible. Depending on how many steps of unit time ahead we want to predict, we distinguish one-step, two-step, or m-step prediction. The more steps we take, or the further in the future we want to look, the larger the prediction error will be. We deal here only with one-step prediction.

The signal model is again a first-order autoregressive process, described in section 7.2,

$$x(k) = ax(k-1) + w(k-1) \tag{7.24}$$

and the observation (or measurement) is affected by additive white noise, i.e.

$$y(k) = cx(k) + v(k) \tag{7.25}$$

We want the 'best' linear estimate of $x(k+1)$, i.e. the signal at time $k+1$, given the data and previous estimate at time k. We denote this one-step prediction estimate as $x(k+1|k)$. By 'best' we mean the predictor that minimizes the mean-square prediction error

$$p(k+1|k) = E[e^2(k+1|k)]$$
$$= E[x(k+1) - \hat{x}(k+1|k)]^2 \tag{7.26}$$

This corresponds to the mean-square error

$$p(k) = E[x(k) - \hat{x}(k)]^2$$

in the filtering case. Strictly, a filtered estimate should be denoted as $\hat{x}(k|k)$.

For a one-step linear predictor, we choose the recursive form used earlier, i.e.

$$\hat{x}(k+1|k) = \alpha(k)\hat{x}(k|k-1) + \beta(k)y(k) \tag{7.27}$$

The parameters $\alpha(k)$ and $\beta(k)$ are determined from the minimization of the mean-square prediction error given by equation 7.26. Substituting equation 7.27 into equation 7.26 and differentiating, we obtain a set of orthogonal equations similar to those derived in the previous section

$$E[e(k+1|k)\hat{x}(k|k-1)] = 0 \tag{7.28}$$

$$E[e(k+1|k)y(k)] = 0 \tag{7.29}$$

The relationship between $\alpha(k)$ and $\beta(k)$,

$$\alpha(k) = a - c\beta(k) \tag{7.30}$$

is determined, from equation 7.28, in a similar way to the relationship between $a(k)$ and $b(k)$ is derived in section 8 of the appendix for the filter case. Substituting this result into the prediction equation, we have

$$\hat{x}(k+1|k) = a\hat{x}(k|k-1) + \beta(k)[y(k) - c\hat{x}(k|k-1)] \tag{7.31}$$

The parameter $\beta(k)$ is determined, together with $p(k+1|k)$, from equations 7.29 and 7.26. Using a similar method to that used in section 9 of the appendix, we obtain

$$p(k+1|k) = \frac{a}{c}\sigma_v^2\beta(k) + \sigma_w^2 \tag{7.32}$$

where $$\beta(k) = \frac{acp(k|k-1)}{c^2 p(k|k-1) + \sigma_v^2} \tag{7.33}$$

Equation 7.33 enables us to calculate $\beta(k)$ from the previous mean-square predictor error, and equation 7.32 then gives us the mean-square predictor error for $p(k+1|k)$.

As before, the optimum processor multiplies the previous estimate by a, and then adds a weighted correction term. Note that the correction term consists of the exact difference between the new data sample $y(k)$ and the previous prediction estimate $c\hat{x}(k|k-1)$. In the filtering problem considered in section 7.3, the correction term involved $y(k)$ minus a times the previous estimate, as seen by comparing equation 7.31 with equation 7.17.

Assuming that the random driving force in equation 7.24 is zero, the signal evolves according to the equation $x(k) = ax(k-1)$. Therefore, given an estimate $\hat{x}(k)$ at time k, it seems reasonable to predict the estimate at time $k+1$ as

$$\hat{x}(k+1|k) = a\hat{x}(k) \tag{7.34}$$

when no other information is available. It can be shown that this intuitive form of the estimate is also valid for the model driven by the noise $w(k-1)$. This is because, by hypothesis, the noise $w(k-1)$ is independent of the state at all times earlier than k (see Sorenson(35), pp. 226–8). Now applying equation 7.34 to equation 7.17 we have

$$\begin{aligned} a\hat{x}(k) &= \hat{x}(k+1|k) \\ &= a\hat{x}(k|k-1) + ab(k)[y(k) - c\hat{x}(k|k-1)] \end{aligned} \tag{7.35}$$

which is the same as the predictor equation 7.31, provided that $\beta(k) = ab(k)$, i.e. the prediction gain and filtering gain are also related by the parameter a.

In addition, the mean-square estimation error $p(k)$ and the prediction error $p(k+1|k)$ are related as follows:

$$\begin{aligned} p(k+1|k) &= E[x(k+1) - \hat{x}(k+1|k)]^2 \\ &= E[ax(k) + w(k) - a\hat{x}(k)]^2 \\ &= E\{a[x(k) - \hat{x}(k)] + w(k)\}^2 \end{aligned}$$

where we have used equations 7.24 and 7.34. Since $w(k)$ is not correlated with the error term $x(k) - \hat{x}(k)$, we have

$$p(k+1|k) = a^2 p(k) + \sigma_w^2 \qquad (7.36)$$

Applying this to equation 7.32, with equation 7.33 substituted for $\beta(k)$, we obtain the mean-square estimation error $p(k)$ which is the same as $p(k)$ from equations 7.18 and 7.19 derived in section 7.3.

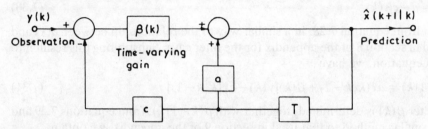

Fig. 7.6   Optimum recursive one-step predictor

The optimum one-step prediction is shown in fig. 7.6, and optimum filtering and prediction simultaneously are shown in fig. 7.7. Solutions for the one-step predictor are given by equations 7.31 to 7.33 but, we need these equations in a form suitable for direct transformation to vector signals. Equations 7.31 and 7.33 are already in this form, but equation 7.32 must be modified using equation 7.33 to eliminate σ_v^2.

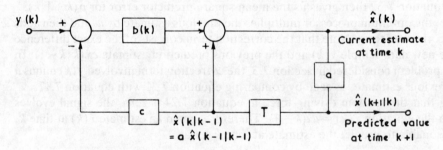

Fig. 7.7   Filtering and prediction simultaneously

> *Predictor equation:*
> $$\hat{x}(k+1|k) = a\hat{x}(k|k-1) + \beta(k)[y(k) - c\hat{x}(k|k-1)] \qquad (7.37)$$
>
> *Predictor gain:*
> $$\beta(k) = acp(k|k-1)[c^2 p(k|k-1) + \sigma_v^2]^{-1} \qquad (7.38)$$
>
> *Prediction mean-square error:*
> $$p(k+1|k) = a^2 p(k|k-1) - ac\beta(k)p(k|k-1) + \sigma_w^2 \qquad (7.39)$$

We shall refer to the above set as the *scalar Kalman predictor* for the signal model

$$x(k+1) = ax(k) + w(k) \qquad (7.40)$$

and the measurement or observation model

$$y(k) = cx(k) + v(k) \qquad (7.41)$$

We return to these equations in section 8.4.

7.5   Comments on Kalman filter

Similarly to Wiener filters discussed in section 6.4, we examine here the frequency domain behaviour of the Kalman filter. For this purpose we set $b(k) = b$, in equation 7.20, and rewrite it as

$$\hat{x}(k) = by(k) + a(1 - bc)\hat{x}(k - 1)$$

where we shall consider b as a parameter under control of the set of equations 7.21 to 7.23.

Applying z-transform, we can form the following transfer function

$$H_b(z) = \frac{bz}{z - a(1 - bc)} \tag{7.42}$$

This is a first order function, similar to the one in section 6.4, with the pole (in z-plane) given by

$$z = a(1 - bc) \tag{7.43}$$

We examine this pole movement, within the unit circle, for two extremes: no observation noise ($\sigma_v^2 = 0$), and very large observation noise ($\sigma_v^2 \to \infty$).

In the first case ($\sigma_v^2 = 0$), equation 7.18 gives $b = 1/c$, while for the case $\sigma_v^2 \to \infty$, we have $b \to 0$. This means that in the first case, the pole is at origin and the transfer function 7.42 is an all-pass, i.e. it is frequency independent. It passes the signal without any restriction.

However, if the noise is very large, we have $b \to 0$, so the pole position is then given by the signal parameter a. If a is close to unity then the transfer function 7.42 becomes narrowband, i.e. it is highly selective, suppressing the noise (as shown in fig. 6.2).

As mentioned earlier, the value of b is not fixed, but it is a function of the Kalman filter algorithm, equations 7.21 to 7.23. Its value controls the pole position and hence the cutoff frequency of the Kalman filter. In fact, the Kalman filter is seen to be a lowpass filter with an *adaptive cutoff* frequency.

It is very similar to the Wiener IIR filter where the pole, given here by a parameter b (different from b in the Kalman filter case), results from the partitioning of $S_{yy}(z)$. This parameter is a function of the signal-to-noise ratio, as pointed out in section 6.4, but it is fixed for a given value of the signal-to-noise. When this changes, new b is obtained, set to this value, etc. So, although Wiener IIR filter solution gives a recursive (and hence sequential) processor, its cutoff frequency is fixed.

On the other hand the Kalman filter which is chosen from the start to be a recursive process, equation 7.11), is therefore a sequential processor but it is also adaptive. Its *adaptation is automatic* and follows the algorithm given by equations 7.21 to 7.23.

8

Optimum estimation of vector signals

8.0   Introduction

In this chapter we deal with vector or multidimensional signals. It is shown in section 8.1 how vector equations are formulated in the case of simultaneous estimation of a number of signals, or in the case of signals generated by higher-order systems. These vector equations result in matrix operations on vectors, so, as discussed in section 8.2, the estimation problem for multidimensional systems is formulated in terms of vectors and matrices. These are of the same form as the scalar equations in chapter 7, so, since there is an equivalence between scalar and matrix operations, all results in chapter 7 for scalar signals are transformed into vector and matrix equations in sections 8.3 and 8.4.

8.1   Signal and data vectors

We have dealt so far with scalar random signals generated by a first-order autoregressive process. We want to extend the same type of representation to broader classes of signals, and also to simultaneous estimation of several signals (for example, components of three-dimensional position and velocity vectors). It is shown below that these multi-dimensional signals are conveniently represented by vector notation, and in place of simple gain parameters we then have matrix operations on vectors. To illustrate the formation of vector equations we consider several examples (Schwartz & Shaw(27), p. 340).

Example 8.1
Assuming that we have q independent signals to be estimated or predicted *simultaneously*, we denote samples of these signals, at time k, as $x_1(k)$, $x_2(k)$, ..., $x_q(k)$. Assuming also that each one is generated by its own first-order autoregressive process, the αth signal sample, for example, is formed according to the equation

$$x_\alpha(k) = a_\alpha x_\alpha(k-1) + w_\alpha(k-1) \qquad \text{where } \alpha = 1, 2, \ldots, q \tag{8.1}$$

Each of the w_α processes is assumed to be white, zero-mean and independent of all others. We define the q-dimensional vectors made up of the q signals and q white noise driving processes as

114

$$\mathbf{x}(k) = \begin{bmatrix} x_1(k) \\ x_2(k) \\ \cdot \\ \cdot \\ \cdot \\ x_q(k) \end{bmatrix} \qquad \text{and } \mathbf{w}(k) = \begin{bmatrix} w_1(k) \\ w_2(k) \\ \cdot \\ \cdot \\ \cdot \\ w_q(k) \end{bmatrix} \qquad (8.2)$$

In terms of these defined vectors, the q equations 8.1 can be written as the *first-order vector equation*

$$\mathbf{x}(k) = A\mathbf{x}(k-1) + \mathbf{w}(k-1) \qquad (8.3)$$

where $\mathbf{x}(k)$, $\mathbf{x}(k-1)$and $\mathbf{w}(k-1)$ are $(q \times 1)$ column vectors and A is a $(q \times q)$ matrix, diagonal in this case, given by

$$A = \begin{bmatrix} a_1 & 0 & . & . & 0 \\ 0 & a_2 & . & . & 0 \\ \cdot & & & & \\ \cdot & & & & \\ \cdot & & & & \\ 0 & . & . & . & a_q \end{bmatrix} \qquad (8.4)$$

Example 8.2
Let the signal $x(k)$ obey not a first-order but a second-order recursive difference equation

$$x(k) = ax(k-1) + bx(k-2) + w(k-1) \qquad (8.5)$$

Such an equation can often arise in a real system, describing its dynamical behaviour, or it may fit a measured set of data better than the first-order difference equation.

To transform equation 8.5 into the first-order vector equation we define two components

$$x_1(k) = x(k) \qquad \text{and} \qquad x_2(k) = x(k-1) = x_1(k-1)$$

Now equation 8.5 can be written as two equations

$$\left. \begin{array}{l} x_1(k) = ax_1(k-1) + bx_2(k-1) + w(k-1) \\ x_2(k) = x_1(k-1) \end{array} \right\} \qquad (8.6)$$

where the first equation is equation 8.5 rewritten using defined components (or states), and the second equation represents the relationship between defined components.

Forming the two-dimensional vector

$$\mathbf{x}(k) = \begin{bmatrix} x_1(k) \\ x_2(k) \end{bmatrix}$$

we can combine the two equations 8.6 into the single vector equation

$$\underbrace{\begin{bmatrix} x_1(k) \\ x_2(k) \end{bmatrix}}_{\mathbf{x}(k)} = \underbrace{\begin{bmatrix} a & b \\ 1 & 0 \end{bmatrix}}_{A} \underbrace{\begin{bmatrix} x_1(k-1) \\ x_2(k-1) \end{bmatrix}}_{\mathbf{x}(k-1)} + \underbrace{\begin{bmatrix} w(k-1) \\ 0 \end{bmatrix}}_{\mathbf{w}(k-1)} \qquad (8.7)$$

This is again in the form of the first-order vector equation.

Example 8.3

We consider a radar tracking problem, to be treated in more detail in chapter 9. Assume a vehicle being tracked is at range $R + \rho(k)$ at time k, and at range $R + \rho(k+1)$ at time $k+1$, T seconds later. We use T to represent the spacing between samples made one scan apart. The average range is denoted by R, and $\rho(k)$, $\rho(k+1)$ represent deviations from the average. We are interested in estimating these deviations, which are assumed to be statistically random with zero-mean value.

To a first approximation, if the vehicle is travelling at radial velocity $\dot{\rho}(k)$ and T is not too large

$$\rho(k+1) = \rho(k) + T\dot{\rho}(k) \tag{8.8}$$

which is the range equation, as given for range prediction in section 1.2.

Similarly, considering acceleration $u(k)$ we have

$$Tu(k) = \dot{\rho}(k+1) - \dot{\rho}(k) \tag{8.9}$$

which is the acceleration equation. Assuming that $u(k)$ is a zero-mean, stationary white noise process, the acceleration is, on average, zero and uncorrelated between intervals, i.e. $E[u(k+1)u(k)] = 0$, but it has some known variance $E[u^2(k)] = \sigma_u^2$. Such accelerations might be caused by sudden wind gusts or short-term irregularities in engine thrust. The quantity $u_1(k) = Tu(k)$ is also a white noise process, and we have in place of equation 8.9

$$\dot{\rho}(k+1) = \dot{\rho}(k) + u_1(k) \tag{8.10}$$

We define now a two component signal vector $\mathbf{x}(k)$ with one component the range, $x_1(k) = \rho(k)$, and the other component the radial velocity, $x_2(k) = \dot{\rho}(k)$. Applying these to equations 8.8 and 8.10, we have

$$x_1(k+1) = x_1(k) + Tx_2(k)$$
$$x_2(k+1) = x_2(k) + u_1(k)$$

or combining them into a single vector equation, we obtain

$$\underbrace{\begin{bmatrix} x_1(k+1) \\ x_2(k+1) \end{bmatrix}}_{\mathbf{x}(k+1)} = \underbrace{\begin{bmatrix} 1 & T \\ 0 & 1 \end{bmatrix}}_{A} \underbrace{\begin{bmatrix} x_1(k) \\ x_2(k) \end{bmatrix}}_{\mathbf{x}(k)} + \underbrace{\begin{bmatrix} 0 \\ u_1(k) \end{bmatrix}}_{\mathbf{w}(k)} \tag{8.11}$$

which is again in the form of the first-order vector equation 8.3.

Equations with time-varying coefficients can be handled as well by defining a time-varying matrix $A(k, k-1)$, also known as the system transition matrix; see, for example, Sage & Melsa(36).

8.1.1   Data vector

Assume that in estimating the signal vector $\mathbf{x}(k)$ we made r *simultaneous noisy measurements at time* k. These measurement samples are labelled $y_1(k)$, $y_2(k)$, ..., $y_r(k)$, so we have the following set of data.

$$y_1(k) = c_1 x_1(k) + v_1(k)$$
$$y_2(k) = c_2 x_2(k) + v_2(k)$$

$$\cdot$$
$$\cdot$$
$$\cdot$$

$$(8.12)$$

$$y_r(k) = c_r x_r(k) + v_r(k)$$

where $v_j(k)$ terms represent additive noise and $c_1, \ldots, c_r$ are some measurement parameters, similar to c introduced in equation 7.9. This set of equations can be put into vector form by defining r-component vectors $\mathbf{y}(k)$ and $\mathbf{v}(k)$. In terms of the previously defined q-component signal vector $\mathbf{x}(k)$, we have the *data vector*:

$$\mathbf{y}(k) = C\mathbf{x}(k) + \mathbf{v}(k) \qquad (8.13)$$

where $\mathbf{y}(k)$ and $\mathbf{v}(k)$ are $(r \times 1)$ column vectors, $\mathbf{x}(k)$ is a $q \times 1)$ column vector, and C is an $(r \times q)$ observation matrix, which in this case, assuming $r < q$, is given by

$$
C = \;
r \Bigg\downarrow
\overset{\displaystyle \xrightarrow{\hspace{3cm}} q}{
\begin{bmatrix}
c_1 & 0 & . & . & 0 & . & . & 0 \\
0 & c_2 & & & & & & . \\
. & & . & & & & & . \\
. & & & . & & & & . \\
0 & . & & & c_r & & & 0
\end{bmatrix}}
\qquad (8.14)
$$

Example 8.4
If $y_1(k) = 4x_1(k) + 5x_2(k) + v_1(k)$, and all the other coefficients in equation 8.12 are equal to unity, we have

$$y_1(k) = 4x_1(k) + 5x_2(k) + 0x_3(k) + \ \ldots \ + v_1(k)$$
$$y_2(k) = 0x_1(k) + \ x_2(k) + 0x_3(k) + \ \ldots \ + v_2(k)$$

$$\cdot$$
$$\cdot$$
$$\cdot$$

$$y_r(k) = 0x_1(k) + \ldots \ldots \ldots \quad + x_r(k) + v_r(k)$$

In this case the observation matrix is

$$
C =
\begin{bmatrix}
4 & 5 & 0 & . & . & . & . & 0 \\
0 & 1 & 0 & & & & & . \\
0 & 0 & 1 & & & & & . \\
. & & & & & & & . \\
. & & & & & & & \\
0 & . & . & . & 1 & . & . & 0
\end{bmatrix}
$$

Example 8.5
For the previous radar tracking case, the signals to be estimated are the range $\rho(k)$, the radial velocity $\dot{\rho}(k)$, the bearing (azimuth) $\theta(k)$, and the angular velocity $\dot{\theta}(k)$. Then $\mathbf{x}(k)$ is the four-component vector

$$\mathbf{x}(k) = \begin{bmatrix} \rho(k) \\ \dot\rho(k) \\ \theta(k) \\ \dot\theta(k) \end{bmatrix}$$

However, measurements of range and bearing only are made, in the presence of additive noise $v_1(k)$ and $v_2(k)$ respectively. In this case we have $q = 4$ and $r = 2$, and the velocities are then found in terms of these quantities, using equations such as 8.8. The matrix C is given here by

$$C = \begin{bmatrix} 1 & 0 & 0 & 0 \\ 0 & 0 & 1 & 0 \end{bmatrix}$$

The block diagram for the system and measurements in vector forms is the same as for the scalar case, figs 7.3(a) and (b), respectively. However, now the notation changes to vectors, and the system and observation parameters become matrices, as in figs 8.1 and 8.2. Here and in other figures shown later, vectors are denoted by underbars.

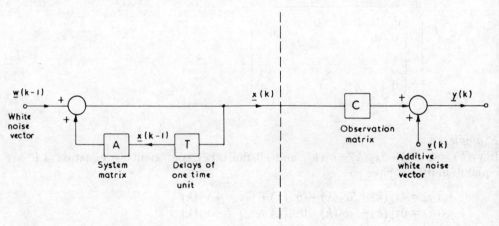

Fig. 8.1   System model   **Fig. 8.2**   Measurement (observation) model

8.2   Vector problem formulation

Returning to the basic problem, we have a signal vector $\mathbf{x}(k)$ obeying a known first-order vector dynamical equation

$$\mathbf{x}(k+1) = A\mathbf{x}(k) + \mathbf{w}(k) \tag{8.15}$$

to be extracted from a noisy measurement vector $\mathbf{y}(k)$

$$\mathbf{y}(k) = C\mathbf{x}(k) + \mathbf{v}(k) \tag{8.16}$$

These two vector equations are obtained as discussed in section 8.1.

The problem is how to form $\hat{\mathbf{x}}(k)$, the 'best' linear estimate (filtered value) of $\mathbf{x}(k)$, and how to form $\hat{\mathbf{x}}(k\,|\,k-1)$, the 'best' predicted value. By 'best' we now mean estimators that minimize the mean-square error of each signal component simultaneously. For example, in the filtering operation each mean-square error

$$E[x_\alpha(k) - \hat{x}_\alpha(k)]^2 \qquad \text{where } \alpha = 1, 2, \ldots, q \qquad (8.17)$$

is to be minimized.

The problem is formally the same as that stated previously in dealing with single time-varying signals obeying a first-order dynamical equation. In the multidimensional case, or the more complex signal case, we could reformulate all the equations in vector form and apply a matrix minimization procedure to obtain the optimum solutions. This procedure would be very similar to the ones used for the single signal case, but now repeated in vector form. However, since we already have the solutions for the one-dimensional (scalar) cases we can extend them to the multidimensional (vector) systems, using the equivalence of scalar and matrix operations in Table 8.1, in which the superscript T stands for transpose of a matrix, and -1 for the inverse of a matrix.

Table 8.1 Transformation of scalar to matrix

Scalar	$\rightarrow$	Matrix
$a + b$		$A + B$
ab		AB
$a^2 b$		ABA^T
$\dfrac{1}{a + b}$		$(A + B)^{-1}$

We have already seen, in section 8.1, that in transition from the single signal to vector signal, the system parameter a changed into the system matrix A, and the data coefficient c changed into the observation matrix C. We now consider the transition of other relevant quantities.

The transition from the observation noise variance to the observation noise covariance matrix (common variance of a number of signals) is written as

$$\sigma_v^2 = \sigma_{v1,1}^2 = E[v_1^2(k)] \rightarrow R(k) = E[\mathbf{v}(k)\mathbf{v}^T(k)] \qquad (8.18)$$

where we have used the third entry of Table 8.1, with $b = 1$. For example, for two signals, we have

$$R(k) = \begin{bmatrix} E[v_1^2(k)] & E[v_1(k)v_2(k)] \\ E[v_2(k)v_1(k)] & E[v_2^2(k)] \end{bmatrix}$$

$$= \begin{bmatrix} \sigma_{v1,1}^2 & \sigma_{v1,2}^2 \\ \sigma_{v2,1}^2 & \sigma_{v2,2}^2 \end{bmatrix}$$

Similarly, for the system noise, we have

$$\sigma_w^2 = \sigma_{w1,1}^2 = E[w_1^2(k)] \rightarrow Q(k) = E[\mathbf{w}(k)\mathbf{w}^T(k)] \qquad (8.19)$$

where $Q(k)$ represents the system noise covariance matrix. If there is no correlation between noise processes, the off-diagonal terms are zero.

The mean-square error for the single signal changes into the error covariance matrix

$$p(k) = p_{1,1}(k) = E[e_1^2(k)] \rightarrow P(k) = E[\mathbf{e}(k)\mathbf{e}^T(k)] \qquad (8.20)$$

For two signals, we have

$$P(k) = \begin{bmatrix} E[e_1^2(k)] & E[e_1(k)e_2(k)] \\ E[e_2(k)e_1(k)] & E[e_2^2(k)] \end{bmatrix}$$

$$= \begin{bmatrix} p_{1,1}(k) & p_{1,2}(k) \\ p_{2,1}(k) & p_{2,2}(k) \end{bmatrix} \tag{8.21}$$

Note that the diagonal terms are the individual mean-square errors as formulated by equation 8.17. They can also be expressed as $E[e^T(k)e(k)]$ which are sums of variances (= trace of covariance matrix).

8.3   Vector Kalman filter

We are now in a position to transform the scalar Kalman filter algorithm, given by equations 7.20 to 7.23, into the corresponding vector Kalman filter. With reference to these equations, and discussions in sections 8.1 and 8.2, we can write directly the vector and matrix equations, tabulated below.

Estimator:
$$\hat{\mathbf{x}}(k) = A\hat{\mathbf{x}}(k-1) + K(k)[\mathbf{y}(k) - CA\hat{\mathbf{x}}(k-1)] \tag{8.22}$$

Filter gain:
$$K(k) = P_1(k)C^T[CP_1(k)C^T + R(k)]^{-1} \tag{8.23}$$

$$\text{where} \quad P_1(k) = AP(k-1)A^T + Q(k-1) \tag{8.24}$$

Error covariance matrix:
$$P(k) = P_1(k) - K(k)C(k)P_1(k) \tag{8.25}$$

The above equations constitute the *vector Kalman filter* for the model described by the state equations

$$\mathbf{x}(k) = A\mathbf{x}(k-1) + \mathbf{w}(k-1) \tag{8.26}$$
$$\mathbf{y}(k) = C\mathbf{x}(k) + \mathbf{v}(k) \tag{8.27}$$

introduced and discussed in section 8.1 (equations 8.3 and 8.13).

According to the scalar-matrix equivalence, Table 8.1, it would have been more correct to write $B(k)$ in place of $b(k)$. We have used $K(k)$ instead, since this is a commonly used notation for the gain matrix in the Kalman filter. Other quantities have been arranged, within the rules of Table 8.1, to obtain the standard form of Kalman equations used, ref. (35), (36), (37). It will be found that some authors use the notation $P(k|k-1)$ and $P(k|k)$ in place of $P_1(k)$ and $P(k)$ respectively. The quantity $P(k|k-1)$ is then referred to as the *predicted covariance matrix*. The notation $P_1(k)$ and $P(k)$, based on a similar one used by Sorenson(36), has been chosen here for simplicity. In equation 8.24, we have $Q(k-1)$ since σ_w^2 in fact represents $E[w^2(k-1)]$, as shown in section 9 of the appendix. As mentioned earlier in section 8.1, for time-varying systems and time-varying observations, matrices A and C are functions of time, i.e. we have $A(k, k-1)$ and $C(k)$ respectively.

To check that equations 8.22 to 8.25 are written correctly, we write the corresponding dimensional equations:

$$q \times 1 = (q \times q)(q \times 1) + (q \times r)[(r \times 1) - (r \times q)(q \times q)(q \times 1)] \tag{8.22$'$}$$
$$q \times r = (q \times q)(q \times r)[(r \times q)(q \times q)(q \times r) + (r \times r)]^{-1} \tag{8.23$'$}$$

Similar equations can be written for equations 8.24 and 8.25. This is in general a very useful method of checking whether vector or matrix equations are written correctly.

Eliminating $K(k)$ from equation 8.25 by means of equation 8.23, changing $P(k)$ to $P(k-1)$, and substituting it into equation 8.24, we obtain a nonlinear difference equation. This equation is known as the *Ricatti* difference equation whose solution in a closed form is generally complex. However, if a numerical procedure of the Kalman algorithm (i.e. equations 8.22 and 8.25) is carried out, there is no need to have a closed form solution for $P_1(k)$.

One of the most significant features of the Kalman filter is its recursive form property that makes it extremely useful in processing measurements to obtain the optimal estimate, utilizing a digital computer. The measurements may be processed as they occur, and it is not necessary to store any measurement data. Only $\hat{x}(k-1)$ need to be stored in proceeding from time $(k-1)$ to time k. However, the algorithm does require storage of the time histories of the matrices $A(k)$ and $C(k)$ if they are time-varying, $Q(k-1)$, and $R(k)$ for all $k = 1, 2, \ldots$. These fall into two types: first, the physical model must be defined (A, C), and second, the statistics of the random processes must be known (Q, R).

The information flow in the filter can be discussed very simply by considering the block diagram fig. 8.3, which is a representation of equation 8.22. Let us suppose that $\hat{x}(k-1)$

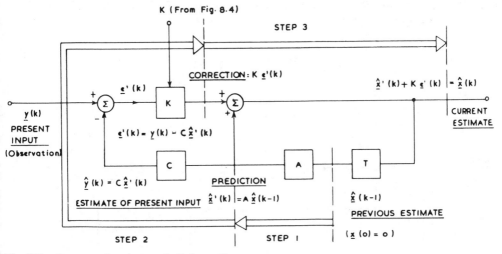

Fig. 8.3 Computational steps in Kalman filter

is known for some k and that we seek to determine $\hat{x}(k)$, given $y(k)$. The computational cycle would proceed as follows:

(i) the estimate $\hat{x}(k-1)$ is 'propagated forward' by premultiplying it by the system matrix A. This gives the predicted estimate $\hat{x}(k|k-1)$, denoted as $\hat{x}'(k)$;

(ii) $\hat{x}'(k)$ is premultiplied by C giving $\hat{y}(k)$ which is subtracted from the actual measurement $y(k)$ to obtain the measured residual (or error) $e'(k)$;

(iii) the residual is premultiplied by the matrix $K(k)$ and the result is added to $x'(k)$ to give $\hat{x}(k)$;

(iv) $\hat{x}(k)$ is stored until the next measurement is made, at which time the cycle is repeated.

The filter operates in a 'predict-correct' fashion, i.e. the 'correction' term $K(k)\mathbf{e}'(k)$ is added to the predicted estimate $\hat{\mathbf{x}}'(k)$ to determine the filtered estimate. The correction term involves the Kalman gain matrix $K(k)$. A similar discussion is given in section 7.4 in the paragraph below equation 7.33. In order to initiate filtering, we start with $\hat{\mathbf{x}}(0) = 0$, and we see immediately that $\hat{\mathbf{x}}(1) = K(1)\mathbf{y}(1)$. Then $\hat{\mathbf{x}}(2), \hat{\mathbf{x}}(3), \ldots$ follow recursively, as discussed in the above four steps.

It is interesting to note that the optimal filter shown in fig. 8.3 consists of the *model of the dynamic process*, which performs the function of prediction, and a feedback correction scheme in which the product of gain and the residual term is applied to the model as a forcing function (compare this system with fig. 8.1). The same comment applies to the scalar Kalman filter shown in fig. 7.5 and the model in fig. 7.3.

The gain matrix $K(k)$ can be calculated before estimation is carried out since it does not depend at all on the measurements. This approach requires storing the calculated vectors for each recursion and feeding them out as needed. However, in the subroutine algorithm given by equations 8.23 and 8.25, in which gains are updated recursively as the estimation proceeds, there is no need to store all gain values: the previous values is the only one required. In this case the subroutine computational diagram is shown in fig. 8.4, and the computational cycle would proceed as follows:

(i)   given $P(k-1)$, $Q(k-1)$, $A(k, k-1)$, then $P_1(k)$ is computed using equation 8.24;
(ii) $P_1(k)$, $C(k)$ and $R(k)$ are substituted into equation 8.23 to obtain $K(k)$, which is used in step three of the filter computation given earlier;

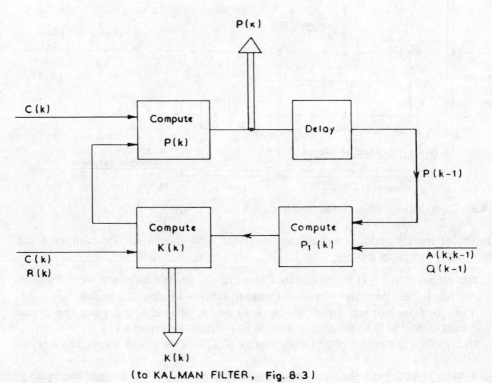

Fig. 8.4 Subroutine calculations for Kalman filter

(iii) $P_1(k)$, $K(k)$ and $C(k)$ are substituted into equation 8.25 to determine $P(k)$, which is stored until the time of the next measurement, when the cycle is repeated.

The matrix inverse which must be computed in equation 8.23 is generally no problem. The matrix to be inverted is $(r \times r)$, as shown in equation 8.23', where r is the number of elements in the measurement vector (see equation 8.13). In most systems r is kept small to avoid the high cost of complex instrumentation. Consequently, it is not unusual to encounter systems with 12 to 15 state variables (in vector $\mathbf{x}$), but only two to three measurement variables (vector $\mathbf{y}$). If r is large, it will also be computationally costly to perform the matrix inversion, and the alternative then is to apply the matrix inversion lemma.

Equations 8.23 and 8.25 define the algorithm for the recursive computation of the optimal filter gain matrix $K(k)$. At the same time, we obtain values for $P_1(k)$ and $P(k)$, i.e. the variances of the components of the prediction and filtering errors respectively.

It is useful for calculations and also for better understanding to group the results for the Kalman filter, equations 8.22 to 8.25, into the following two stages:

(i)   *Prediction*

$$\hat{\mathbf{x}}(k|k-1) = A\hat{\mathbf{x}}(k-1|k-1) \tag{8.28}$$

$$P(k|k-1) = AP(k-1|k-1)A^{\mathrm{T}} + Q(k-1) \tag{8.29}$$

(ii)   *Updating (or correction)*

$$\hat{\mathbf{x}}(k|k) = \hat{\mathbf{x}}(k|k-1) + K[\mathbf{y}(k) - C\hat{\mathbf{x}}(k|k-1)] \tag{8.30}$$

$$P(k|k) = P(k|k-1) - KCP(k|k-1) \tag{8.31}$$

$$K = P(k|k-1)C^{\mathrm{T}}[CP(k|k-1)C^{\mathrm{T}} + R(k)]^{-1} \tag{8.32}$$

where $P(k|k-1)$ and $P(k|k)$ correspond to $P_1(k)$ and $P(k)$ respectively in the Kalman filter given by equations 8.23 to 8.25. The first stage is prediction based on the state equation 8.26, and the second stage is the updating or correction based on the measurement equation 8.27; these stages are also illustrated in fig. 8.5. This presentation

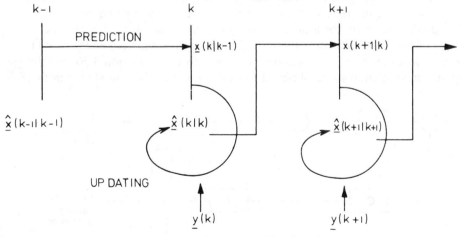

Fig. 8.5   Illustration for two stage Kalman computational cycle

of the Kalman filter is in a form particularly suitable for computations, but the earlier form given by equations 8.22 to 8.25 is more suitable as an introduction.

8.4   Vector Kalman predictor

We already have the scalar Kalman predictor algorithm, equations 7.37 to 7.39, arranged in a form suitable for transformation into vector equations. Using the same kind of reasoning as in sections 8.1 to 8.3, we obtain the following vector and matrix set of equations.

Predictor equation:
$$\hat{x}(k+1|k) = A\hat{x}(k|k-1) + G(k)[y(k) - C\hat{x}(k|k-1)] \qquad (8.33)$$

Predictor gain:
$$G(k) = AP(k|k-1)C^{T}[CP(k|k-1)C^{T} + R(k)]^{-1} \qquad (8.34)$$

Prediction mean-square error:
$$P(k+1|k) = [A - G(k)C]P(k|k-1)A^{T} + Q(k) \qquad (8.35)$$

These equations represent the *vector Kalman predictor* for the model described earlier by equations 8.26 and 8.27. We have introduced here $G(k)$ as the predictor gain matrix with its scalar value denoted in section 7.4 as $\beta(k)$. Other quantities are the same as the ones used in the previous section.

As in the scalar case given by equation 7.34, and by Sorenson(36), we have again the following interesting connection between estimated and predicted signal vectors

$$\hat{x}(k+1|k) = A\hat{x}(k) \qquad (8.36)$$

This means that given the filtered signal $\hat{x}(k)$, the best estimate of the signal one step in the future ignores noise and assumes that the signal dynamics matrix A operates only on the estimate. Multiplying the estimator (filter) equation 8.22 by A and using equation 8.36, we obtain the predictor equation 8.33, with $G(k) = AK(k)$. This is the relationship corresponding to the scalar relationship $\beta(k) = ab(k)$ discussed in section 7.4 below equation 7.35. Since in the prediction case we are interested in the prediction error covariance, the three matrix equations 8.23 to 8.25 reduce to two equations 8.34 and 8.35. The prediction error covariance matrix given by equation 8.35 is obtained after substitution of equation 8.25 into equation 8.24 and expressing $P_1(k)$ as $P(k|k-1)$.

The simple link between filtering and prediction, given by equation 8.36, enables the one-step vector prediction to be obtained from the Kalman filter, as shown in fig. 8.6.

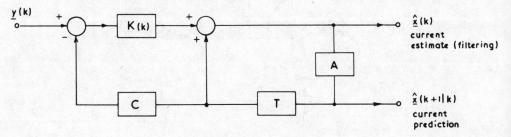

Fig. 8.6   Simultaneous filtering and prediction of vector signals

This is, in fact, the vector case corresponding to fig. 7.7. Note that the A box and the time-delay element are interchanged in this version of the Kalman filter.

8.5   Practical aspects of Kalman filtering

The current estimate in fig. 8.3 is given by

$$\hat{\mathbf{x}}(k) = \hat{\mathbf{x}}'(k) + K\mathbf{e}'(k) \tag{8.37}$$

where $\hat{\mathbf{x}}'(k)$ is the predicted estimate and

$$\mathbf{e}'(k) = \mathbf{y}(k) - C\mathbf{x}'(k) \tag{8.38}$$

is the measured residual (or error), often referred to as the *innovation sequence*. The predicted part $\mathbf{x}'(k)$ is a function of the state-space model (A), and the innovation (or new information) $\mathbf{e}'(k)$ is a function mainly of the new measurement. We observe that the estimator operation is pivoted about the values of the gain or weighting matrix K. For small K, the estimator 'believes' the model, and for large K, the estimator 'believes' the measurement.

This can be confirmed by examining the Kalman gain expressed in an alternative way (see example 8.5) as

$$K(k) = P(k)C^{T}R^{-1}(k) \tag{8.39}$$

where $P(k)$ is the estimation error covariance matrix and $R(k)$ the observation noise covariance matrix.

One can observe now that for small $P(k)$, i.e. good system model, and large $R(k)$ for the use of bad measurement, $K(k)$ will be small. This means the estimator 'believes' the model. On the other hand, for large $P(k)$, i.e. bad model, and small $R(k)$ for the case of good measurement, the estimator 'believes' the measurement.

Furthermore, one can deduce, Candy(38), that K can be expressed as

$$K \propto \frac{Q}{R} \tag{8.40}$$

where Q and R, as defined in section 8.2, refer to system and observation noise respectively. Considering the Kalman estimator as a deterministic filter one can deduce that as Q increases, K increases, and the filter bandwidth increases. The same effect is achieved by small R. On the other hand if Q decreases (R increases), K decreases which reduces the bandwidth. These points are similar to comments on Kalman filters made in section 7.5, where the scalar Kalman gain has been denoted as b.

Example 8.5
Derivation of equation 8.39, by using the *matrix inversion lemma*:

$$(A + BCD)^{-1} = A^{-1} - A^{-1}B(DA^{-1}B + C^{-1})^{-1}DA^{-1} \tag{8.41}$$

Substitute 8.23 into 8.25 and use the above lemma to obtain:

$$P = P_1 - P_1 C^{T}(CP_1 C^{T} + R)^{-1}CP = (P_1^{-1} + C^{T}R^{-1}C)^{-1} \tag{8.42}$$

Insert $I = PP^{-1}$ and $I = R^{-1}R$, into equation 8.23, and obtain

$$K = PP^{-1}P_1C^TR^{-1}R(CP_1C^TR^{-1}R+R)^{-1}$$
$$= PP^{-1}P_1C^TR^{-1}(CP_1C^TR^{-1}+I)^{-1}$$

Now substituting for P^{-1}, from equation 8.42, we have

$$K = PC^TR^{-1}(C^TR^{-1}CP_1+I)(CP_1C^TR^{-1}+I)^{-1} = PC^TR^{-1}$$

A Kalman filter does not function correctly when K becomes small, but the measurements still contain information for the estimates. In such conditions the filter is said to *diverge*. It is then necessary to determine how well the filter functions and how to adjust (or 'tune' it) if necessary. When the Kalman filter is 'tuned' it has optimal performance.

The innovation sequence, $e'(k)$, provides the starting point for checking the filter operation. A necessary and sufficient condition for a Kalman filter to be optimal is that the innovation sequence, $e'(k)$, is zero-mean and white. For this purpose various statistical tests need to be done as discussed and illustrated by simulations of some case studies in Candy(38).

As discussed in section 8.3, in order to apply a Kalman filter to a specific system, it is necessary to specify the system model in terms of parameters (A, C), noise statistics (Q, R) and initial conditions $(x(0), P(0))$. However, such a model is often an approximation to the actual physical system, and the model parameters and noise statistics are rarely exact. The modelling errors or 'model mismatches' may even cause the Kalman filter to diverge. It can be shown that the innovations sequence is no longer zero-mean and white when such model mismatch occurs, Candy(38).

The Kalman filter theory developed in this book is only an introduction. A more realistic picture is obtained by considering possible malfunctioning situations briefly discussed in this section.

Furthermore, the Kalman filter introduced here is the linear discrete time Kalman filter. This algorithm can be modified for the continuous-discrete estimation problems. The other modification may be needed for applications in which the noise sources are correlated.

Also, the linear filter can be used to solve system identification and deconvolution problems. Finally, the linear Kalman filter can be extended to approximate a solution to the nonlinear filtering problem. More information on topics discussed in this section can be found in Candy(38).

9

Examples

9.0 Introduction

The notation in part 2 such as x, $\hat{x}$, y, P, Q, R, K is the same as that generally used in the literature on Kalman filtering, but the notation used in state equations is different. We are using the discrete-time state equations written as

$$
\begin{aligned}
\mathbf{x}(k) &= A\mathbf{x}(k-1) + \mathbf{w}(k-1) \\
\mathbf{y}(k) &= C\mathbf{x}(k) + \mathbf{v}(k)
\end{aligned}
\tag{9.1}
$$

where A and C, if time-varying, would be denoted respectively as $A(k, k-1)$ $C(k)$. This notation is analogous to the continuous-time state equations written as

$$
\begin{aligned}
\dot{\mathbf{x}}(t) &= A(t)\mathbf{x}(t) + \mathbf{w}(t) \\
\mathbf{y}(t) &= C(t)\mathbf{x}(t)
\end{aligned}
\tag{9.2}
$$

The alternative notation to the one used in equations 9.1 is

$$
\begin{aligned}
\mathbf{x}(k) &= \Phi(k, k-1)\mathbf{x}(k-1) + \mathbf{w}(k-1) \\
\mathbf{y}(k) &= H(k)\mathbf{x}(k) + \mathbf{v}(k)
\end{aligned}
\tag{9.3}
$$

where $\Phi(k, k-1)$, sometimes denoted as $F(k)$, is the state transition matrix corresponding to $A(k, k-1)$ in equation 9.1, and $H(k)$ is the observation matrix corresponding to $C(k)$ in equation 9.1. The notation of equation 9.3 will be found in most literature on Kalman filtering (or general estimation theory), for example in references (27, 33, 36, 37). However, it is interesting to note that in some text books (for example Freeman(6), p. 19), A, B, C notation is used for the discrete-time systems in the same way as in this book, where the additional letter B is the matrix relating the control input to the state vector, which has not been used in this book except later in examples 9.4 and 9.6.

There are seven examples in this chapter. The first three are standard ones illustrating the basic manipulation techniques for Wiener and Kalman filters. The others are more advanced, dealing with various Kalman filter applications.

9.1 Scalar Wiener filter

Estimate the random amplitude x of a sinusoidal waveform of known frequency ω, in the presence of additive noise $v(t)$, using the scalar Wiener filter. The measurement is represented as

$$
y(t) = x \cos \omega t + v(t)
\tag{9.4}
$$

where the additive noise $v(t)$ accounts for both receiver noise and inaccuracies of the instruments used. Let $y(t)$ be sampled at $\omega t = 0$ and $\omega t = \pi/4$, giving us measured data $y(1) = y(\omega t = 0)$ and $y(2) = y(\omega t = \pi/4)$. Assume that $E(x) = 0$, $E(x^2) = \sigma_x^2$, $E[v(1)v(2)] = 0$, $E[v(1)]^2 = E[v(2)]^2 = \sigma_v^2$, and also that x is independent of v.

The linear estimator for this case is given by equation 6.7 with optimum weights expressed by equation 6.14 or 6.15. We have two measurement samples $y(1) = x + v_1$ and $y(2) = (x/\sqrt{2}) + v_2$. Therefore, expanding equation 6.14 for $i = j = 1, 2$, we have

$$h_0(1)p_{yy}(1, 1) + h_0(2)p_{yy}(2, 1) = p_{xy}(1)$$
$$h_0(1)p_{yy}(1, 2) + h_0(2)p_{yy}(2, 2) = p_{xy}(2) \tag{9.5}$$

where $p_y(i, j)$ and $p_{xy}(j)$ are given by equations 6.12 and 6.13 respectively. Solving equations 9.5 for $h(1)$ and $h(2)$, and substituting the values for $p_y(i, j)$ and $p_{xy}(j)$, we have

$$h_0(1) = \frac{\sigma_x^2}{\frac{3}{2}\sigma_x^2 + \sigma_v^2} \tag{9.6}$$

$$h_0(2) = \frac{h_0(1)}{\sqrt{2}} \tag{9.7}$$

Using these solutions in equation 6.7 we obtain the estimate of x as

$$\hat{x} = \frac{\sigma_x^2}{\frac{3}{2}\sigma_x^2 + \sigma_v^2} y(1) + \frac{1}{\sqrt{2}} \frac{\sigma_x^2}{\frac{3}{2}\sigma_x^2 + \sigma_v^2} y(2) \tag{9.8}$$

where $y(1)$ and $y(2)$ are the measurement samples.

9.2 Scalar Kalman filter

Estimate the value of a constant x, given discrete measurements of x corrupted by an uncorrelated Gaussian noise sequence with zero mean and variance σ_v^2.

The scalar equations describing this situation are

$$x(k) = x(k-1)$$

for the system and

$$y(k) = x(k) + v(k)$$

for the measurement, where

$$E[v(k)] = 0, \qquad E[v^2(k)] = \sigma_v^2$$

In this problem $a = 1$, and $w(k-1) = 0$, therefore $p_1(k) = p(k-1)$, see equation 7.22. Also, $c = 1$, so that the filter (or Kalman) gain, from equation 7.21 is given by

$$b(k) = \frac{p(k-1)}{p(k-1) + \sigma_v^2} \tag{9.9}$$

The error covariance, given by equation 7.23, in this case becomes

$$p(k) = \frac{p(k-1)}{1 + p(k-1)/\sigma_v^2}$$

which is a difference equation. This equation can be solved by noting that

$$p(1) = \frac{p(0)}{1 + p(0)/\sigma_v^2}$$

$$p(2) = \frac{p(1)}{1 + p(1)/\sigma_v^2} = \frac{p(0)}{1 + 2p(0)/\sigma_v^2}$$

$$\vdots$$

$$p(n) = \frac{p(0)}{1 + np(0)/\sigma_v^2} \tag{9.10}$$

Having obtained the above solution, we use it to calculate

$$b(k) = \frac{p(0)}{\sigma_v^2 + kp(0)} \tag{9.11}$$

Now we can write for the optimum (Kalman) discrete filter, described by equation 7.20, for this example:

$$\hat{x}(k) = \hat{x}(k-1) + \frac{p(k-1)}{\sigma_v^2 + p(k-1)}[y(k) - \hat{x}(k-1)] \tag{9.12}$$

where we have used equation 9.9 for b. From the above result we see that without measurement noise, i.e. $\sigma_v^2 = 0$, we have $\hat{x}(k) = y(k)$. But with noise, it is better to use equation 9.11 for b, in which case the Kalman filter is given by

$$\hat{x}(k) = \hat{x}(k-1) + \frac{p(0)}{\sigma_v^2 + kp(0)}[y(k) - \hat{x}(k-1)] \tag{9.13}$$

For sufficiently large k, we have $\hat{x}(k) = \hat{x}(k-1) = \hat{x}$, indicating that further measurements provide no new information.

9.3   Vector Kalman filter

Consider the second-order message model given by

$$\mathbf{x}(k) = \begin{bmatrix} 1 & 1 \\ 0 & 1 \end{bmatrix} \mathbf{x}(k-1) + \mathbf{w}(k-1)$$

The state is observed by means of a scalar observation model

$$y(k) = x_1(k) + v(k) \tag{9.14}$$

The input noise is stationary with

$$Q(k) = Q \begin{bmatrix} 0 & 0 \\ 0 & 1 \end{bmatrix}$$

and the measurement noise is non-stationary, with $R(k) = 2 + (-1)^k$. In other words, the measurements for even values of k have more noise than the odd values of k. We assume that the variance of the initial errors (or initial state) is given by

$$P(0) = \begin{bmatrix} 10 & 0 \\ 0 & 10 \end{bmatrix}$$

We wish to compute the values of $K(k)$ for $k = 1$ to 10.
We start with the covariance equation 8.24, i.e.

$$P_1(k) = AP(k-1)A^T + Q(k-1)$$

which for this problem becomes

$$P_1(k) = \begin{bmatrix} 1 & 1 \\ 0 & 1 \end{bmatrix} P(k-1) \begin{bmatrix} 1 & 0 \\ 1 & 1 \end{bmatrix} + \begin{bmatrix} 0 & 0 \\ 0 & 1 \end{bmatrix}$$

For $k = 1$, we have

$$P_1(1) = \begin{bmatrix} 1 & 1 \\ 0 & 1 \end{bmatrix} \begin{bmatrix} 10 & 0 \\ 0 & 10 \end{bmatrix} \begin{bmatrix} 1 & 0 \\ 1 & 1 \end{bmatrix} + \begin{bmatrix} 0 & 0 \\ 0 & 1 \end{bmatrix}$$

where we have substituted given initial error covariance $P(0)$. The above results in

$$P_1(1) = \begin{bmatrix} 20 & 10 \\ 10 & 11 \end{bmatrix}$$

which is used in equation 8.23 as follows

$$K(1) = P_1(1)C^T[CP_1(1)C^T + R(1)]^{-1}$$

or
$$K(1) = \begin{bmatrix} 20 & 10 \\ 10 & 11 \end{bmatrix} \begin{bmatrix} 1 \\ 0 \end{bmatrix} \left\{ \begin{bmatrix} 1 & 0 \end{bmatrix} \begin{bmatrix} 20 & 10 \\ 10 & 11 \end{bmatrix} \begin{bmatrix} 1 \\ 0 \end{bmatrix} + [1] \right\}^{-1}$$

$$= \begin{bmatrix} 20 \\ 10 \end{bmatrix} (21)^{-1} = \begin{bmatrix} 20/21 \\ 10/21 \end{bmatrix} = \begin{bmatrix} 0.95 \\ 0.48 \end{bmatrix}$$

giving $K_1(1) = 0.95$ and $K_2(1) = 0.48$.
   Note that in the above calculations we have used

$$c = [1 \quad 0]$$

from the observation equation 9.14, from which we also have for the noise variance

$$R(k) = E[v(k)v^T(k)]$$

$$= E\left\{ \begin{bmatrix} v_1(k) \\ 0 \end{bmatrix} [v_1^T(k) \quad 0] \right\}$$

$$= \begin{bmatrix} E[v_1(k)v_1^T(k)] & \\ & 0 \end{bmatrix} = \begin{bmatrix} R(k) & \\ & 0 \end{bmatrix}$$

where $R(k) = 2 + (-1)^k$.
   To compute $K(2)$, we have first to find the value $P(1)$ from equation 8.25 in the following way:

$$P(k) = [1 - K(k)C]P_1(k)$$

$$P(1) = \left\{ \begin{bmatrix} 1 & 0 \\ 0 & 1 \end{bmatrix} - \begin{bmatrix} 20/21 \\ 10/21 \end{bmatrix} [1 \quad 0] \right\} \begin{bmatrix} 20 & 10 \\ 10 & 11 \end{bmatrix}$$

and after evaluations, we have

$$P(1) = \begin{bmatrix} 0.95 & 0.48 \\ 0.48 & 6.24 \end{bmatrix}$$

Now we use the above value of $P(1)$ to calculate the next value for $P_1(2)$ from

$$P_1(2) = AP(1)A^T + Q$$

$$= \begin{bmatrix} 1 & 1 \\ 0 & 1 \end{bmatrix} \begin{bmatrix} 0.95 & 0.48 \\ 0.48 & 6.24 \end{bmatrix} \begin{bmatrix} 1 & 0 \\ 1 & 1 \end{bmatrix} + \begin{bmatrix} 0 & 0 \\ 0 & 1 \end{bmatrix}$$

$$P_1(2) = \begin{bmatrix} 8.1 & 6.7 \\ 6.6 & 7.2 \end{bmatrix}$$

This result enables us to calculate the set of values for $K(2)$ from

$$K(2) = P_1(2) C^T [C P_1(2) C^T + R(2)]^{-1}$$

$$= \begin{bmatrix} 8.1 & 6.7 \\ 6.6 & 7.2 \end{bmatrix} \begin{bmatrix} 1 \\ 0 \end{bmatrix} \left\{ \begin{bmatrix} 1 & 0 \end{bmatrix} \begin{bmatrix} 8.1 & 6.7 \\ 6.6 & 7.2 \end{bmatrix} \begin{bmatrix} 1 \\ 0 \end{bmatrix} + \begin{bmatrix} 3 \\ 0 \end{bmatrix} \right\}^{-1}$$

$$= \begin{bmatrix} 0.73 \\ 0.6 \end{bmatrix}$$

giving $K_1(2) = 0.73$ and $K_2(2) = 0.6$.

Next we compute $P(2)$ to obtain $P_1(3)$ and then $K(3)$, and so on. The components of $K(k)$ are shown in fig. 9.1 for $k = 1$ to 10. Note how the gain increases for odd values of k, to reflect the less noisy measurements. We see that the gain has reached an approximately periodic steady-state solution after only a few samples.

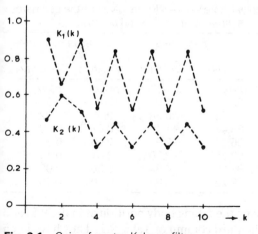

Fig. 9.1   Gain of vector Kalman filter

9.4   Kalman filter application to falling body

Consider a noise-free second-order system representing a falling body in a constant field

$$\ddot{z} = -g \qquad t \geq 0 \tag{9.15}$$

Let the position be $z = x_1$, and velocity $\dot{z} = x_2$.

Then, defining the vector

$$\mathbf{x}(t) = \begin{bmatrix} x_1(t) \\ x_2(t) \end{bmatrix}$$

equations 9.15 can be written in state-space form

$$\dot{\mathbf{x}}(t) = F\mathbf{x}(t) + Gu(t) \tag{9.16}$$

where $F = \begin{bmatrix} 0 & 1 \\ 0 & 0 \end{bmatrix}$, $G = \begin{bmatrix} 0 \\ 1 \end{bmatrix}$, and $u(t) = -g$ is a constant. For the Kalman filter application, we need to obtain the difference equation

$$\mathbf{x}(kT) = A\mathbf{x}[(k-1)T] + Bu(t) \tag{9.17}$$

From appendix 12, we have for this case

$$A = e^{FT} = I + \begin{bmatrix} 0 & 1 \\ 0 & 0 \end{bmatrix} T + \begin{bmatrix} 0 & 1 \\ 0 & 0 \end{bmatrix}^2 \frac{T^2}{2} = \begin{bmatrix} 1 & T \\ 0 & 1 \end{bmatrix}$$

and $\quad B = \left(\int_0^T e^{F\tau} d\tau \right) g = \int_0^T \begin{bmatrix} 1 & \tau \\ 0 & 1 \end{bmatrix} \begin{bmatrix} 0 \\ 1 \end{bmatrix} d\tau = \begin{bmatrix} T^2/2 \\ T \end{bmatrix}$ $\tag{9.18}$

With these values for A and B, and for $T = 1$, we have

$$\mathbf{x}(k) = \underbrace{\begin{bmatrix} 1 & 1 \\ 0 & 1 \end{bmatrix}}_{A} \mathbf{x}(k-1) + \underbrace{\begin{bmatrix} 1/2 \\ 1 \end{bmatrix}}_{B} \underbrace{(-g)}_{u} \tag{9.19}$$

which is the difference equation representing our system, i.e. a falling body in a constant field.

Table 9.1 Falling body in a constant field. The initial state values $\hat{x}(0)$ are given as the estimates at $t = kT = 0$, at which time the assumed errors in estimates are $p_{11}(0)$ and $p_{22}(0)$

	True values		Position observations	Estimates		Errors in estimates	
	Position	Velocity		Position	Velocity	Position	Velocity
$t = kT$	$x_1(t)$	$x_2(t)$	$y(k)$	$\hat{x}_1(k)$	$\hat{x}_2(k)$	$p_{11}(k)$	$p_{22}(k)$
0	100.0	0		95.0	1.0	10.0	1.0
1	99.5	−1.0	100.0	99.63	0.38	0.92	0.92
2	98.0	−2.0	97.9	98.43	−1.16	0.67	0.58
3	95.5	−3.0	94.4	95.21	−2.91	0.66	0.30
4	92.0	−4.0	92.7	92.35	−3.70	0.61	0.15
5	87.5	−5.0	87.3	87.68	−4.84	0.55	0.08
6	82.0	−6.0	82.1	82.22	−5.88	0.50	0.05

This system has been observed by measuring the falling body position as in fig. 9.2, and the results of measurements are shown in the third column of Table 9.1. They have been affected by some independent random disturbance $v(k)$, so that the observation equation, in this case, can be written as

$$y(k) = \underbrace{[1 \quad 0]}_{C} \mathbf{x}(k) + v(k) \tag{9.20}$$

with the random disturbance (or noise) variance given by $R(k) = \sigma_v^2 = 1$.

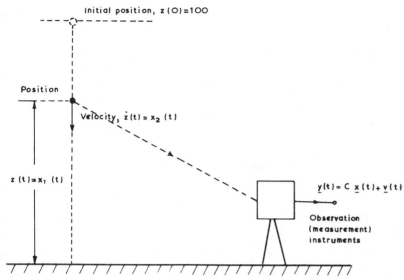

Fig. 9.2   Measurements of falling body for application of Kalman filter

Furthermore, we assume that the initial value of the state vector (position and velocity) is

$$\hat{\mathbf{x}}(0) = \begin{bmatrix} 95 \\ 1 \end{bmatrix}$$

and $P(0) = \begin{bmatrix} 10 & 0 \\ 0 & 1 \end{bmatrix}$

are the assumed errors at time $k = 0$. Matrices A and C are indicated in equations 9.19 and 9.20. It will be noticed that we have, in this case, another term denoted by $B\mathbf{u}$. This is the driving force term which has been left out in our system equation 8.15, but we can easily incorporate it in the Kalman algorithm because it changes only the predictor term in equations 8.22 and 8.28, see also fig. 8.3, in the following way:

$$\hat{\mathbf{x}}'(k) = A\hat{\mathbf{x}}(k-1) + B\mathbf{u} \tag{9.21}$$

Therefore, the Kalman filter equation (8.22), for this case, is written as

$$\hat{\mathbf{x}}(k) = \hat{\mathbf{x}}'(k) + K(k)[y(k) - C\hat{\mathbf{x}}'(k)] \tag{9.22}$$

where $\hat{\mathbf{x}}'(k)$ is given by equation 9.21. All other equations for $P_1(k)$, $K(k)$ and $P(k)$ are unaltered. (Analysis of model with deterministic and random forcing functions can be found, for example, in Sorenson(35), pp. 226–9.)

To calculate the state estimates we start with equation 8.24, with $Q = 0$ for this case, and proceed as follows

$$P_1(1) = AP(0)A^{\mathrm{T}} = \begin{bmatrix} 1 & 1 \\ 0 & 1 \end{bmatrix}\begin{bmatrix} 10 & 0 \\ 0 & 1 \end{bmatrix}\begin{bmatrix} 1 & 0 \\ 1 & 1 \end{bmatrix}$$

$$= \begin{bmatrix} 11 & 1 \\ 1 & 1 \end{bmatrix}$$

Then, using equation 8.23, with $P_1(1)$ given by the above result, C as in equation 9.20, and $R(k) = 1$, we have

$$K(1) = P_1(1)C^T[CP_1(1)C^T + R]^{-1}$$

$$= \begin{bmatrix} 11 & 1 \\ 1 & 1 \end{bmatrix}\begin{bmatrix} 1 \\ 0 \end{bmatrix}\left\{[1 \quad 0]\begin{bmatrix} 11 & 1 \\ 1 & 1 \end{bmatrix}\begin{bmatrix} 1 \\ 0 \end{bmatrix} + 1\right\}^{-1}$$

$$= \begin{bmatrix} 11/12 \\ 1/12 \end{bmatrix}$$

Next we calculate the prediction term in equation 9.22, i.e.

$$\hat{x}'(1) = A\hat{x}(0) + Bu$$

$$= \begin{bmatrix} 1 & 1 \\ 0 & 1 \end{bmatrix}\begin{bmatrix} 95 \\ 1 \end{bmatrix} + \begin{bmatrix} -0.5 \\ -1 \end{bmatrix}$$

$$= \begin{bmatrix} 95.5 \\ 0 \end{bmatrix}$$

where we have used $\hat{x}(0)$ specified earlier and Bu term from equation 9.19 with $g = 1$.

Now we have all the components required in equation 9.22, which gives us the state estimate (or filtering equation) as

$$\hat{x}(1) = \hat{x}'(1) + K(1)[y(1) - C\hat{x}'(1)]$$

$$= \begin{bmatrix} 95.5 \\ 0 \end{bmatrix} + \begin{bmatrix} 11/12 \\ 1/12 \end{bmatrix}\left\{100 - [1 \quad 0]\begin{bmatrix} 95.5 \\ 0 \end{bmatrix}\right\}$$

or $\qquad \hat{x}(1) = \begin{bmatrix} 99.6 \\ 0.37 \end{bmatrix}$

therefore $\hat{x}_1(1) = 99.6$ and $\hat{x}_2(1) = 0.37$. This is the first set of values estimated from Kalman filtering and entered in Table 9.1. The errors of the above estimated pair of values are calculated from equation 8.25 as follows:

$$P(1) = P_1(1) - K(1)CP_1(1)$$

$$= \begin{bmatrix} 11 & 1 \\ 1 & 1 \end{bmatrix} - \begin{bmatrix} 11/12 \\ 1/12 \end{bmatrix}[1 \quad 0]\begin{bmatrix} 11 & 1 \\ 1 & 1 \end{bmatrix}$$

$$= \begin{bmatrix} 11/12 & 1/12 \\ 1/12 & 11/12 \end{bmatrix}$$

The diagonal terms $p_{11}(1) = p_{22}(1) = 11/12$ are the ones of interest to us, because they represent the mean-square errors of position and velocity estimates. These have been entered in Table 9.1 for $k = 1$. The first computational cycle has now been completed. The results obtained $P(1)$ and $x(1)$, are then used to compute $x(2)$ and $P(2)$, and so on, until we reach $x(6)$ and $P(6)$. The computer program is in section 10 of the appendix with comments; it has been used to calculate estimates and errors. The results obtained are shown in Table 9.1, and in fig. 9.3.

In this problem we are observing the position, and as a result, the position error drops fast as soon as the first observation is processed. However, the velocity error does not

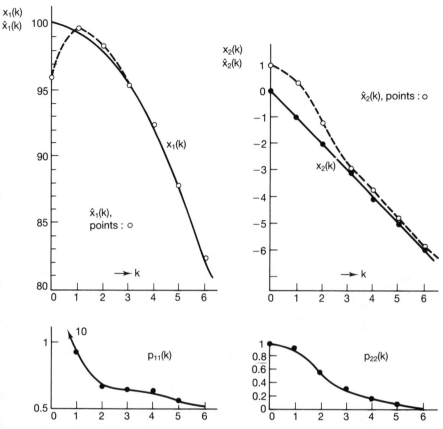

Fig. 9.3   Graphs for table 9.1

decrease much until the second observation is processed, because two position observations are required to determine both components of the state vector. The system is such that velocity affects the position, but position does not affect the velocity. Therefore, the velocity must first be estimated accurately before good estimates of position can be obtained. The above, and also the effect of the initial error conditions on the filtering procedure, are discussed by Jazwinski(37).

9.5   Scalar Kalman filter formulation for RC circuit

We consider the voltage measurement at the output of the RC circuit in fig. 9.4, using a high-impedance voltmeter. Because these measurements are noisy, and also the component values imprecise ($\pm\Delta R, \pm\Delta C$), we require an improved estimate of the output voltage. For this purpose we want to use a Kalman filter for which we develop the system and measurement models as follows.

The Kirchoff nodal equation, for this circuit, is

$$u(t) - \frac{x(t)}{R} - C\frac{dx(t)}{dt} = 0 \tag{9.23}$$

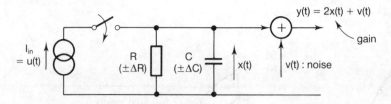

Fig. 9.4　Circuit for example 9.5

with the assumption that the initial voltage across the capacitor is $x(0)$. We rewrite this differential equation in the usual form as

$$\dot{x}(t) = fx(t) + gu(t) \qquad (9.24)$$

where $f = -(1/RC)$, $g = 1/C$, and $\dot{x}(t) = \mathrm{d}x/\mathrm{d}t$.

For the discrete-time Kalman filter application we need to obtain the difference equation based on equation 9.24. One can approach this in two ways as discussed in appendix 11. We choose the simpler one given by (A.10) resulting in the system (or signal) model equation

$$x(kT) = \left(1 - \frac{T}{RC}\right)x[(k-1)T] + \frac{T}{C}u[(k-1)T] \qquad (9.25)$$

The measurement equation, indicated in fig. 9.4, can be written directly in the discrete-time form as

$$y(kT) = 2x(kT) + v(kT) \qquad (9.26)$$

Assuming the circuit elements to have values $R = 3.3\,\mathrm{k\Omega}$, $C = 1000\,\mu\mathrm{F}$ and sampling period $T = 0.1\,\mathrm{s}$, our signal and measurement equations are

$$x(k) = 0.97x(k-1) + 100u(k-1) + w(k-1) \qquad (9.27)$$
$$y(k) = 2x(k) + v(k) \qquad (9.28)$$

In the above, the time has been scaled to $0.1\,\mathrm{s}$ as the unit, and $w(k-1)$ has been added representing the model noise, as used in section 8.1. Assuming the input to be a step function of $300\,\mu\mathrm{A}$, the second term on the right-hand side in equation 9.27 becomes 0.03. The model parameter uncertainty is expressed in terms of the variance σ_w^2 (assumed to be 0.0001), and the measurement errors in terms of the variance σ_v^2 (assumed to be 4).

For filtering we are given $(a,\ b,\ c,\ \sigma_v^2,\ x(0),\ p(0))$, the input $\{u(k)\}$, noisy measurement $\{y(k)\}$ sequence, and the algorithm (equations 7.20 to 7.23). For this example, the computational algorithm is written as follows:

$$\hat{x}(k|k-1) = 0.97\hat{x}(k-1|k-1) + 0.03 \qquad (9.29)$$
$$p(k|k-1) = 0.94p(k-1|k-1) + 0.0001 \qquad (9.30)$$

$$e'(k) = y(k) - 2\hat{x}(k|k-1) \qquad (9.31)$$
$$Pe'_{(k)} = 4p(k|k-1) + 4 \qquad (9.32)$$

$$b(k) = \frac{p(k|k-1)}{2[p(k|k-1) + 1]} \qquad (9.33)$$
$$\hat{x}(k|k) = \hat{x}(k|k-1) + b(k)e(k) \qquad (9.34)$$

$$p(k|k) = \frac{p(k|k-1)}{p(k|k-1)+1} = 2b(k) \tag{9.35}$$

The above sequence of equations is written in the computational form indicated by equations 8.28 to 8.32. This means the first pair (equations 9.29 and 9.30) are the state prediction and prediction error, while the last three are updating equations, after the measurement $y(k)$. However, an additional feature is also given by the pair (equations 9.31 and 9.32) which are the innovation (introduced in section 8.5) and its error. Such an arrangement gives us the information about correct functioning of the algorithm.

Having set the model equations 9.27 and 9.28, and the filter estimator algorithm (equations 9.29 to 9.35), we need to perform a sequence of computations for $k = 1, 2, 3,$ .... For this purpose one needs computer support preferably as a software package. Results for the case discussed here can be seen in Candy, Ch. 5(38).

9.6   Vector Kalman filter formulation for RLC circuit

In this case we consider the design of an estimator for a second-order system consisting of R, L, C elements, fig. 9.5. The loop equation for this circuit is

$$\frac{d^2x(t)}{dt^2} + \frac{R}{L}\frac{dx(t)}{dt} + \frac{1}{LC}x(t) = \frac{1}{LC}u(t) \tag{9.36}$$

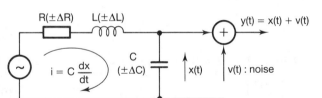

Fig. 9.5   Circuit for example 9.6

where $u(t)$ is the forcing signal. Assuming $R = 5\,\text{k}\Omega$, $L = 2.5\,\text{H}$, and $C = 0.1\,\mu\text{F}$, we have $R/L = 2\times10^3$, $1/LC = 4\times10^6$, and if we scale time from seconds to milliseconds we obtain 10^6 as a common factor which can be eliminated, hence simplifying equation 9.36 to

$$\frac{d^2x(t)}{dt^2} + 2\frac{dx(t)}{dt} + 4x(t) = 4u(t) \tag{9.37}$$

The above dynamic equation can be placed into the state-space form, similarly as in section 8.1, by defining states $x_1 = x$, $x_2 = dx/dt$. Hence, we have

$$\left.\begin{aligned}\frac{dx_1}{dt} &= x_2 \\[2mm] \text{and} \quad \frac{dx_2}{dt} &= -4x_1 - 2x_2 + 4u\end{aligned}\right\} \tag{9.38}$$

where the second state equation follows from equation 9.37. This set of equations can be written in matrix form as

$$\dot{\mathbf{x}}(t) = F\mathbf{x}(t) + Gu(t) + Gw(t) \tag{9.39}$$

where $F = \begin{bmatrix} 0 & 1 \\ -4 & -2 \end{bmatrix}$ and $G = \begin{bmatrix} 0 \\ 4 \end{bmatrix}.$ $\tag{9.40}$

The additional term $Gw(t)$, in equation 9.39, is a random component due to model component inaccuracies; G applies to both $u(t)$ and $w(t)$ because they are located in the same place in the circuit.

The measurement, fig. 9.5, is given by

$$y(t) = x(t) + v(t) \tag{9.41}$$

The circuit model equation 9.39 and the measurement model equation 9.41 are in the continuous-time form. Before we design the discrete Kalman estimator, we need to convert these two models to the discrete-time (sampled-data) form. The discrete-time form for equation 9.41, with $t = kT$, is simply

$$y(kT) = x(kT) + v(kT)$$
or $\qquad y(k) = C\mathbf{x}(k) + v(k)$ $\tag{9.42}$

where $C = [1 \quad 0]$, and taking T as unity time interval.

On the other hand, the discrete-time form for equation 9.39 is

$$\mathbf{x}(k) = A\mathbf{x}(k-1) + Bu(k-1) + Bw(k-1) \tag{9.43}$$

where $\quad A = e^{FT} = I + FT + \dfrac{(FT)^2}{2!} + \dots$ $\tag{9.44}$

and $\quad B = F^{-1}(e^{FT} - I)G$ $\tag{9.45}$

For F and G given in equation 9.40, and for $T = 0.1$ ms, we have

$$A = \begin{bmatrix} 0.98 & 0.09 \\ -0.36 & 0.8 \end{bmatrix}, B = \begin{bmatrix} 0.02 \\ 0.36 \end{bmatrix} \tag{9.46}$$

where A has been obtained using only three terms in equation 9.44. Having established the signal (or system) and measurement models we can now formulate the corresponding vector Kalman filter algorithm as follows

Prediction:
$\hat{x}(k|k-1) = A\hat{x}(k-1|k-1) + Bu(k-1)$ (state prediction)
$P(k|k-1) = AP(k-1)|k-1)A^{\mathrm{T}} + Q(k-1)$ (prediction covariance)

Innovation:
$e'(k) = y(k) - \hat{y}(k|k-1) = y(k) - C\hat{x}(k|k-1)$ (innovation)
$R_{e(k)} = CP(k|k-1)C^{\mathrm{T}} + R$ (innovation covariance)

Gain:
$K(k) = P(k|k-1)CR_e^{-1}(k)$ (Kalman gain or weight)

Correction:
$\hat{x}(k|k) = \hat{x}(k|k-1) + K(k)e'(k)$ (state correction)
$P(k|k) = [I - K(k)C]P(k|k-1)$ (correction covariance)

Initial conditions:
$\hat{x}(0|0), P(0,0)$

In this example, Q and R are

$$Q = \begin{bmatrix} \sigma_w^2 & 0 \\ 0 & \sigma_w^2 \end{bmatrix}, \quad R = \begin{bmatrix} \sigma_v^2 & 0 \\ 0 & \sigma_v^2 \end{bmatrix}$$

Note too that in the state prediction we have also the forcing term, introduced earlier in example 9.4, equation 9.21. Results of a computer simulation of this example, for $u(t)$ being a pulse train excitation, can be found in Candy, Ch. 5(38).

9.7   Kalman filter formulation for radar tracking

This example shows what must be done prior to the application of the Kalman filter algorithm to radar tracking. A brief introduction is given in the paragraph below.

In radar tracking, the time delay between transmission and reception of the pulses provides an estimate of the aircraft range (radial distances), while the location of the antenna beam at the time of detection provides the aircraft bearing (azimuth). A short range radar rotates typically at a scan rate of 15 revolutions/minute (r.p.m.), while longer range radars rotate at 6 r.p.m. Therefore, we have for these two cases new range and bearing estimates every 4 s and 10 s respectively. This means that the tracking filters are updated at such an interval, corresponding to the time interval previously denoted by T.

We have already considered a radar tracking case in chapter 8, examples 8.3 and 8.5. In example 8.3 we derived the system model, for variations about average values, with states $x_1(k) = \rho(k)$ for the range and $x_2(k) = \dot{\rho}(k)$ for the radial velocity. We now add two more states concerned with the bearing $x_3(k) = \theta(k)$, and bearing rate (or angular velocity) $x_4(k) = \dot{\theta}(k)$. The addition of these two states augments equation 8.11 into the following system equation:

$$\underbrace{\begin{bmatrix} x_1(k+1) \\ x_2(k+1) \\ x_3(k+1) \\ x_4(k+1) \end{bmatrix}}_{\mathbf{x}(k+1)} = \underbrace{\begin{bmatrix} 1 & T & 0 & 0 \\ 0 & 1 & 0 & 0 \\ 0 & 0 & 1 & T \\ 0 & 0 & 0 & 1 \end{bmatrix}}_{A} \underbrace{\begin{bmatrix} x_1(k) \\ x_2(k) \\ x_3(k) \\ x_4(k) \end{bmatrix}}_{\mathbf{x}(k)} + \underbrace{\begin{bmatrix} 0 \\ u_1(k) \\ 0 \\ u_2(k) \end{bmatrix}}_{\mathbf{w}(k)} \qquad (9.47)$$

In this expression the noise terms $u_1(k)$ and $u_2(k)$ represent the change in radial velocity and bearing rate respectively over interval T. They are each T times radial and angular acceleration. We assume $u_1(k)$ and $u_2(k)$ to be random with zero mean, and we also assume that they are uncorrelated both with each other and individually from one interval to the next.

The radar sensors are assumed to provide noisy estimates of the range $\rho(k) = x_1(k)$, and bearing $\theta(k) = x_3(k)$ at time intervals T. At time k, the two sensor outputs are then

$$y_1(k) = x_1(k) + v_1(k)$$
$$y_2(k) = x_3(k) + v_2(k)$$

Therefore, the data vector, as discussed in chapter 8, equation 8.13 and example 8.5, can be written as

$$
\underbrace{\begin{bmatrix} y_1(k) \\ y_2(k) \end{bmatrix}}_{\mathbf{y}(k)} = \underbrace{\begin{bmatrix} 1 & 0 & 0 & 0 \\ 0 & 0 & 1 & 0 \end{bmatrix}}_{C} \underbrace{\begin{bmatrix} x_1(k) \\ x_2(k) \\ x_3(k) \\ x_4(k) \end{bmatrix}}_{\mathbf{x}(k)} + \underbrace{\begin{bmatrix} v_1(k) \\ v_2(k) \end{bmatrix}}_{\mathbf{v}(k)}
\tag{9.48}
$$

The additive noise, $\mathbf{v}(k)$, is usually assumed to be Gaussian with zero-mean and variances $\sigma_\rho^2(k)$ and $\sigma_\theta^2(k)$. So far, we have established vector equations for the system model given by equation 9.47, and data model given by equation 9.48. The next step is to formulate noise covariance matrices Q for the system, and R for the measurement model. For the latter, using equation 8.18, we have

$$
R(k) = E[\mathbf{v}(k)\mathbf{v}^{\mathrm{T}}(k)] = \begin{bmatrix} \sigma_\rho^2(k) & 0 \\ 0 & \sigma_\theta^2(k) \end{bmatrix}
\tag{9.49}
$$

and the system noise covariance matrix, defined in equation 8.19, is for this case given by

$$
Q(k) = E[\mathbf{w}(k)\mathbf{w}^{\mathrm{T}}(k)] = \begin{bmatrix} 0 & 0 & 0 & 0 \\ 0 & \sigma_1^2 & 0 & 0 \\ 0 & 0 & 0 & 0 \\ 0 & 0 & 0 & \sigma_2^2 \end{bmatrix}
\tag{9.50}
$$

where $\sigma_1^2 = E(u_1^2)$ and $\sigma_2^2 = E(u_2^2)$ are the variances of T times the radial and angular acceleration respectively. Specific values must be substituted for those variances in order to define the Kalman filter numerically. To simplify, we assume that the probability density function (p.d.f) of the acceleration in either direction (ρ or θ) is uniform and equal to $p(u) = 1/2M$, between limits $\pm M$; the variance is $\sigma_u^2 = M^2/3$. A more realistic p.d.f. is used by Schwartz & Shaw(27), based on Singer & Behnke(39). The variance in equation 9.50 are then $\sigma_1^2 = T^2\sigma_u^2$ and $\sigma_2^2 = \sigma_1^2/R^2$.

To start Kalman processing we have to initialize the gain matrix $K(k)$. For this purpose the error covariance matrix $P(k)$ has to be specified in some way. A reasonable *ad hoc* initialization can be established using two measurements, range and bearing, at times $k = 1$ and $k = 2$. From these four measurement data we can make the following estimates:

$$
\hat{\mathbf{x}}(2) = \begin{bmatrix} \hat{x}_1(2) = \hat{\rho}(2) = y_1(2) \\[2mm] \hat{x}_2(2) = \dot{\hat{\rho}}(2) = \dfrac{1}{T}[y_1(2) - y_1(1)] \\[2mm] \hat{x}_3(2) = \hat{\theta}(2) = y_2(2) \\[2mm] \hat{x}_4(2) = \dot{\hat{\theta}}(2) = \dfrac{1}{T}[y_2(2) - y_2(1)] \end{bmatrix}
\tag{9.51}
$$

To calculate $P(2)$, we use the general expression for $P(k)$, equation 8.20, where for $k = 2$, we have

$$
P(2) = E\{[\mathbf{x}(2) - \hat{\mathbf{x}}(2)][\mathbf{x}(2) - \hat{\mathbf{x}}(2)]^{\mathrm{T}}\}
\tag{9.52}
$$

Values for $\hat{x}(2)$ are given by equation 9.51, and using equations 9.47 and 9.48 for $x(2)$, we obtain the following result:

$$x(2) - \hat{x}(2) = \begin{bmatrix} -v_1(2) \\ u_1(1) & -(v_1(2) - v_1(1))/T \\ -v_2(2) \\ u_2(1) & -(v_2(2) - v_2(1))/T \end{bmatrix} \tag{9.53}$$

which is a (4×1) column vector.

In this case the error covariance matrix is a (4×4) matrix

$$P(2) = \begin{bmatrix} p_{11} & p_{12} & p_{13} & p_{14} \\ p_{21} & p_{22} & p_{23} & p_{24} \\ p_{31} & p_{32} & p_{33} & p_{34} \\ p_{41} & p_{42} & p_{43} & p_{44} \end{bmatrix} \tag{9.54}$$

Taking into account the independence of noise sources u and v, and also the independence between individual noise samples, it can be shown that the above matrix simplifies to

$$P(2) = \begin{bmatrix} p_{11} & p_{12} & 0 & 0 \\ p_{21} & p_{22} & 0 & 0 \\ 0 & 0 & p_{33} & p_{34} \\ 0 & 0 & p_{43} & p_{44} \end{bmatrix} \tag{9.55}$$

where
$$\begin{aligned} p_{11} &= \sigma_\rho^2 \\ p_{12} &= p_{21} = \sigma_\rho^2/T \\ p_{22} &= 2\sigma_\rho^2/T^2 + \sigma_1^2 \\ p_{33} &= \sigma_\theta^2 \\ p_{34} &= p_{43} = \sigma_\theta^2/T \\ p_{44} &= 2\sigma_\theta^2/T^2 + \sigma_2^2 \end{aligned} \tag{9.56}$$

As a numerical example, we take for range $R = 160$ km ($= 100$ miles), scan time $T = 15$ s, and a maximum acceleration $M = 2.1$ m s^{-2}. Let the r.m.s. noise in the range sensor be equivalent to 1 km, therefore $\sigma_\rho = 10^3$ m. Furthermore, let r.m.s. noise σ_θ in the bearing sensor be 1° or 0.017 rad. These two figures define numerically the noise covariance matrix R.

For the above numerical values, we can calculate noise variances in the Q matrix, equation 9.50, as $\sigma_1^2 = 330$ and $\sigma_2^2 = 1.3 \times 10^{-8}$.

Using the above values in equation 9.55, we obtain the initial value of the estimation covariance matrix $P(2)$, or in the alternative notation $P(2|2)$,

$$P(2|2) = \begin{bmatrix} 10^6 & 6.7 \times 10^4 & 0 & 0 \\ 6.7 \times 10^4 & 0.9 \times 10^4 & 0 & 0 \\ 0 & 0 & 2.9 \times 10^{-4} & 1.9 \times 10^{-5} \\ 0 & 0 & 1.9 \times 10^{-5} & 2.6 \times 10^{-6} \end{bmatrix} \tag{9.57}$$

Since we have this error matrix at $k = 2$, we could try to use it to calculate the predictor gain $G(3)$ at $k = 3$, which is given by

$$G(3) = AP(3|2)C^{T}[CP(3|2)C^{T}+R]^{-1} \tag{9.58}$$

where all quantities (A, C, R) are known except $P(3|2)$. We might try to use equation 8.35 to calculate $P(3|2)$ as

$$P(3|2) = [A - G(2)C]P(2|1)A^{T} + Q$$

but $G(2)$ and $P(2|1)$ are not known. However, equation 8.35 has been derived from 8.24 and 8.25, so instead of equation 8.35, we can use 8.24 rewritten as

$$P(k|k-1) = AP(k-1|k-1)A^{T} + Q \tag{9.59}$$

which for $k = 3$ becomes

$$P(3|2) = AP(2|2)A^{T} + Q \tag{9.60}$$

where $P(2|2)$ is known from equation 9.57. Substituting for the other matrices from equations 9.47 and 9.50, and using the given numerical values we obtain

$$P(3|2) = \begin{bmatrix} 5 \times 10^{6} & 2 \times 10^{5} & 0 & 0 \\ 2 \times 10^{5} & 9.3 \times 10^{3} & 0 & 0 \\ 0 & 0 & 14.5 \times 10^{-4} & 5.8 \times 10^{-5} \\ 0 & 0 & 5.8 \times 10^{-5} & 2.6 \times 10^{-6} \end{bmatrix} \tag{9.61}$$

The diagonal values gives us the prediction errors. The first and third elements are respectively mean-square range and bearing prediction errors for $k = 3$.

We are now able to calculate the predictor gain $G(3)$ using equation 9.61 in equation 9.58 which gives, after matrix manipulations, the following result:

$$G(3) = \begin{bmatrix} 1.33 & 0 \\ 3.3 \times 10^{-2} & 0 \\ 0 & 1.33 \\ 0 & 3.3 \times 10^{-2} \end{bmatrix} \tag{9.62}$$

The next step is to find $P(3|3)$ using equation 8.25 which for $k = 3$ gives

$$P(3|3) = P(3|2) - K(3)CP(3|2)$$

where $K(3) = A^{-1}G(3)$ (see section 8.4). The process is then repeated by finding $P(4|4)$, $G(4)$ etc. Graphical representation of complete results for this case can be found in Schwartz & Shaw(27), pp. 358–62.

Solutions to problems

Section 1.5

1.2 (a) $F(z) = 1 + \frac{1}{4}z^{-2} + \frac{1}{16}z^{-4}$; (b) $F(z) = \dfrac{1}{1 - \frac{1}{4}z^{-2}}$

1.4 $X(z) = z^2/(z-1)^2$; double zero at 0 and double pole at 1.

1.5 $Y(z)(1 + b_1 z^{-1}) = a_0 X(z)$; $Y(z)(1 - b_1 z^{-1} - b_2 z^{-1}) = a_0 X(z)$

1.7 All have poles at

$$z_{1,2} = (2 - \alpha - \beta)/2 \pm \tfrac{1}{2}(\beta^2 + \alpha^2 + 2\alpha\beta - 4\beta)^{1/2}$$

For critically damped system: $\alpha^2 + 2\alpha\beta + (\beta^2 - 4\beta) = 0$.

1.9 (b) $y = (6, 11, 4)$

Section 2.6

2.1 (a) $[y(k)] = [1, -1.5, 1.25, 1.625, -1.18]$; (b) $[y(k)] = [2, -1, \frac{3}{2}, \frac{7}{4}, -\frac{9}{8}]$

2.4 $y(k) = (-1)^k \dfrac{(1 - b_1^{k+1})}{[1 - b_1]} a_0$

2.5 $[x_1(k)] = [0, 1, 1, 1, 1]$; $[x_2(k)] = [1, \frac{5}{2}, \frac{5}{4}, \frac{5}{8}, \frac{5}{16}]$
$x_1(k) = 1 - \delta(k)$; $x_2(k) = 5(\frac{1}{2})^k - 4\delta(k)$.

2.6 (a) $[x(k)] = [1, 1, \frac{1}{2}, 0, -\frac{1}{4}, -\frac{1}{4}]$; (b) $x(k) = \dfrac{1}{(\sqrt{2})^k}\left(\cos\dfrac{k\pi}{4} + \sin\dfrac{k\pi}{4}\right)$

2.7 (a) $H(z) = \dfrac{0.5z^{-1}}{1 + 0.5z^{-1}}$; (b) $[h(k)] = [0, \frac{1}{2}, -\frac{1}{4}, \frac{1}{8}, -\frac{1}{16}, \frac{1}{32}, \ldots]$

(c) $|H(\omega)| \dfrac{0.45}{(1 + 0.8\cos\omega T)^{1/2}}$ (highpass); $\theta = -\omega T + \tan^{-1}\left(\dfrac{0.5\sin\omega T}{1 + 0.5\cos\omega T}\right)$

2.8 $|H_1(\omega)| = \left|\cos\dfrac{\omega T}{2}\right|$, $\theta_1 = -\omega T/2(z)$; $|H_2(\omega)| = |\cos\omega T|$, $\theta_2 = -\omega T(z \to z^2)$

2.9 $|H_1(\omega)| = \sin(\omega T/2)$, $\theta_1 = (\pi/2) - (\omega T/2)$ (single unit);
$|H_2(\omega)| = \sin^2(\omega T/2)$, $\theta_2 = \pi - \omega T$ (double unit)

2.10 (a) $[y(k)] = [\frac{1}{4}, -\frac{1}{4}, 0, 0, 0, 0]$; (b) $[y(k)] = [0, \frac{1}{4}, 0, 0, 0, 0]$

2.11 (a) $y(k) = \frac{1}{2}[\frac{1}{2}x(k) - x(k-1) + \frac{1}{2}x(k-2)] - \frac{5}{4}y(k-1) - \frac{1}{2}y(k-2)$

(b) $|H(\omega)|^2 = \dfrac{\sin^4(\omega T/2)}{(29/16) + (15/4)\cos\omega T + 2\cos^2\omega T}$

2.12 (a) Double pole at $z_p = 1 - \sqrt{\beta}$. For stability, poles within the unit circle, $0 < \beta < 4$.

(b) $h(k) = (1-k)(0.1)^k$, $(\beta = 0.81)$; $h(k) = (0.2 - 3.6k)(-0.9)^k$, $(\beta = 3.6)$

2.14 $H(z) = 0.447(1 - 1.618z^{-1} + z^{-2})$

Section 3.6

3.1 $[h(k)] = [0, -1, \frac{1}{2}, -\frac{1}{3}, \frac{1}{4}, -\frac{1}{5}, \frac{1}{6}, -\frac{1}{7}, \frac{1}{8}, -\frac{1}{9}]$
$h(-k) = h(k)$

3.2 $|H(\omega)|_3 = 2\sin\omega T$; $|H(\omega)|_5 = 2\sin\omega T - \sin 2\omega T$

3.3 (a) $h(n) = \frac{1}{2}\left(\dfrac{\sin(n\pi/2)}{n\pi/2}\right)$

(b) $[h(k)]_R = \left[\frac{1}{2}, \dfrac{1}{\pi}, 0, -\dfrac{1}{3\pi}, 0, \dfrac{1}{5\pi}, 0\right]$

$[h(k)]_H = \left[\frac{1}{2}, \dfrac{(2-\sqrt{3})\alpha + \sqrt{3}}{2}\left(\dfrac{1}{\pi}\right), 0, \alpha\left(-\dfrac{1}{3\pi}\right), 0, \dfrac{(2\alpha+\sqrt{3})\alpha - \sqrt{3}}{2}\left(\dfrac{1}{5\pi}\right), 0\right]$

3.4 (a) $|H(\omega)| = |0.25 + 0.46\cos\omega T|$;   (b) $|H(\omega)| = |0.25 - 0.46\cos\omega T|$

3.5 (a) $|H(\omega)| = |0.50 - 0.64\cos 2\omega T|$;   (b) $|H(\omega)| = |0.50 + 0.64\cos 2\omega T|$

3.6 $h(n) = (\sin n\pi/2)/n\pi$, $-\infty < n < \infty$
(i) $h_1(n) = h(n)w_R(n)$, $-6 \le n \le 6$:
$h_1(0) = 0.5$, $h_1(\pm 1) = 0.318$, $h_1(\pm 2) = 0$, $h_1(\pm 3) = -0.106$, $h_1(\pm 4) = 0$,
$h_1(\pm 5) = 0.064, h_1(\pm 6) = 0$
(ii) $h_1(0) = 0.54$, $h_1(\pm 1) = 0.318$, $h_1(\pm 2) = -0.04$, $h_1(\pm 3) = -0.108$,
$h_1(\pm 4) = 0.04, h_1(\pm 5) = 0.068, h_1(\pm 6) = -0.05$.

Section 4.5

4.1 (b) $y(k) = (T\sqrt{2}e^{T/\sqrt{2}}\sin(T/\sqrt{2}))x(k-1) + (2e^{-T/\sqrt{2}}\cos(T/\sqrt{2}))y(k-1)$
$- e^{-T\sqrt{2}}y(k-2)$

4.2 (b) $y(k) = \dfrac{T^2}{D}x(k) + \dfrac{2T^2}{D}x(k-1) + \dfrac{T^2}{D}x(k-2) - \dfrac{2T^2-8}{D}y(k-1)$

$- \dfrac{4 - 2\sqrt{2}T + T^2}{D}y(k-2)$

where $D = 4 + 2\sqrt{2}T + T^2$.
As $T \to 0$, the solutions for problems 4.1 and 4.2 reduce to
$$y(k) = T^2x(k-1) + 2y(k-1) - y(k-2)$$

where $x(k) = x(k-1)$, $x(k-2) = x(k-1)$ has been used.

4.3 (a) $\dfrac{\omega_c}{\omega_c + s} \rightarrow \dfrac{\omega_c T}{1 - e^{-\omega_c T} z^{-1}}$; (b) All $\omega_c \rightarrow \omega_c T$.

4.4 (b) $H(z) =$

$$\omega_c^2 T^2 \, \frac{1 + 2z^{-1} + z^{-2}}{[(4 + 2\sqrt{2}\omega_c T + \omega_c^2 T^2) + 2(\omega_c^2 T^2 - 4)z^{-1} + (4 - 2\sqrt{2}\omega_c T + \omega_c^2 T^2)z^{-2}]}$$

4.5 (a) $H(z) = \dfrac{1 - 2z^{-1} + z^{-2}}{(\beta^2 + \beta\sqrt{2} + 1) + (2\beta^2 - 2)z^{-1} + (\beta^2 - \beta\sqrt{2} + 1)z^{-2}}$

H (highpass) $= H$ (lowpass with $z^{-1} \rightarrow -z^{-1}$)

(b) $(2 + \sqrt{2})y(k) = x(k) + x(k-2) - 2x(k-1) - (2 - \sqrt{2})y(k-2)$

4.6 $y(k) = \frac{1}{3}x(k) - \frac{2}{3}x(k-2) + \frac{1}{3}x(k-4) + \sqrt{2}y(k-1) - \frac{2}{3}y(k-2) + \frac{1}{3}\sqrt{2}y(k-3)$
$\quad - \frac{1}{3}y(k-4)$

4.7 $(2 + \sqrt{2})y(k) = [x(k) + x(k-4)] + 4x(k-2) - 2\sqrt{2}[x(k-1) + x(k-3)]$
$\quad\quad\quad + 2(1 + \sqrt{2})y(k-1) - 2y(k-2)$
$\quad\quad\quad + 2(\sqrt{2} - 1)y(k-3) - (2 - \sqrt{2})y(k-4)$

4.10 $\varepsilon_1 = \varepsilon_2 = 0.5\%$ for $\lambda_1 = 200\pi$; $\varepsilon_1 = \varepsilon_2 = 5\%$ for $\lambda_1 = 20\pi$
$\varepsilon_3 = 0.000832\%$ for $\lambda_1 = 200\pi$; $\varepsilon_3 = 0.083\%$ for $\lambda_1 = 20\pi$

Section 5.5

5.1 (a) $F = 4$ Hz; (b) $f_s = 2048$ Hz; (c) f(sig. max.) $= 1024$ Hz

5.2 $F_p(m\Omega) = e^{-jm\pi} 2(1 + \cos(m\pi/2))$

5.3 (a) $F_p(m) = e^{-j(2\pi/5)m} \dfrac{\sin(m\pi/2)}{\sin(m\pi/10)}$; (b) $F_p(m) = \dfrac{\sin(m\pi/2)}{\sin(m\pi/10)}$

5.4 (a) $F(m\Omega) = e^{-j(2\pi/3)m}\left(1 - \cos\left(\dfrac{2\pi}{3}m\right)\right)$, $\Omega = \dfrac{2\pi}{3}$

(b) $F(m\Omega) = e^{-j(2\pi/8)m}\left(1 - \cos\left(\dfrac{2\pi}{8}m\right)\right)$, $\Omega = \dfrac{2\pi}{8}$

5.5 (a) $g^{-1} = h = (\frac{1}{2}, -\frac{1}{4})$; $g^{-1} = h = (\frac{1}{2}, -\frac{1}{4}, \frac{1}{8})$
(b) Error energy: 1/16, for 2-length inverse.
1/64, for 3-length inverse.

5.6 $h_1 = g_1^{-1} = (\frac{1}{2}, -\frac{1}{4}, -\frac{1}{8})$, error energy: 5/32; $h_2 = g_2^{-1} = (1, -1, -1)$, error
energy: 13.
The first sequence, the minimum delay type, has a stable inverse. However, for the
second sequence, the maximum delay type, the inverse is unstable; see, for
example, Robinson(8), p. 156.

Appendix

1 Derivation of second-order difference equation

Applying backward differences as shown by Hovanessian et al.(3) to equation 1.3, we obtain its discrete-time equivalent

$$\frac{y_n - 2y_{n-1} + y_{n-2}}{\Delta t^2} + 2\sigma \frac{y_n - y_{n-1}}{\Delta t} + \omega_0^2 y_n = \omega_0^2 x_n$$

or rewriting this, we have

$$y_n = \frac{2(1 + \sigma\Delta t)}{D} y_{n-1} - \frac{1}{D} y_{n-2} + \frac{\omega_0^2 \Delta t^2}{D} x_n$$

where $D = 1 + 2\sigma\Delta t + \omega_0^2 \Delta t^2$.

In general we can rewrite the above as

$$y_n = b_1 y_{n-1} + b_2 y_{n-2} + a_0 x_n$$

with the coefficients given approximately as follows:

$$b_1 \cong 2(1 + \sigma\Delta t)/\omega_0^2 \Delta t^2$$

$$b_2 \cong \frac{1}{\omega_0^2 \Delta t^2}\left(1 - \frac{2\sigma\Delta t}{1 + \omega_0^2 \Delta t^2}\right) \approx -1/\omega_0^2 \Delta t^2$$

and

$$a_0 = \frac{1}{(1 + 2\sigma\Delta t)/\omega_0^2 \Delta t^2 + 1} \simeq 1 \tag{A.1}$$

since $\omega_0 \gg \sigma$, and assuming $(\omega_0 \Delta t) \gg 1$.

2 Relation between s-plane and z-plane

A point $s_1 = \sigma_1 + j\omega_1$ in the s-plane transforms to a point z_1 in the z plane, defined by equation 1.19, so that we can write

$$z_1 = e^{s_1 T} = e^{\sigma_1 T} e^{j\omega_1 T}$$

This means $|z_1| = e^{\sigma_1 T}$, and $\underline{/z_1} = \omega_1 T$, as shown in fig. A.1(a). Note that the location of point $\underline{/z_1}$ changes as the sampling interval T is altered. To obtain more information on s–z plane relationship, we consider now the case $\sigma_1 = 0$, and find ω_1 for which $\underline{/z_1}$ changes from $-\pi$ to $+\pi$, i.e. $-\pi < \underline{/z_1} < \pi$. Since $\underline{/z_1} = \omega_1 T$, we have further $-\pi < \omega_1 T < \pi$, or $-\pi/T < \omega_1 < \pi/T$, or $-(\omega_s/2) < \omega_1 < (\omega_s/2)$. Therefore, the path along the imaginary axis between $\pm(\omega_s/2)$ transforms into the unit circle in the z-plane, fig. A.1(b), since for $\sigma = 0$, $|z| = 1$. It follows that an l.h.s. strip ω_s wide in the s-plane transforms inside the unit circle ($\sigma < 0$). The corresponding r.h.s. strip ($\sigma > 0$) transforms outside the unit circle as indicated in fig. A.1(c). Successive strips in the s-plane

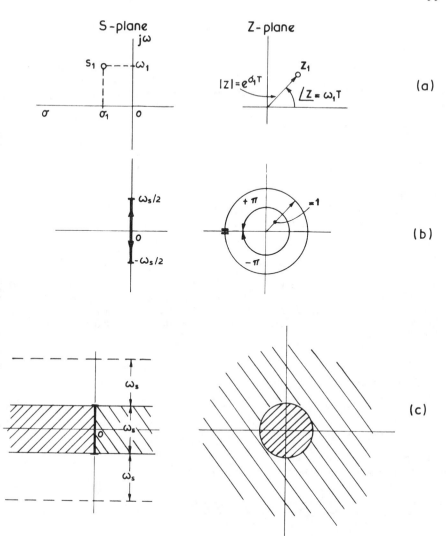

Fig. A.1 (a) Transformation of point in s-plane to point in z-plane; (b) transformation of imaginary axis in s-plane to unit circle in z-plane; (c) transformation of s-plane into z-plane

transform into the same place. Strictly, the strips transform into successive Riemann surfaces superimposed on the z-plane.

Note that the transformation used here, $z = e^{sT}$, is also the one used in control. In communications $z = e^{-sT}$ is used, and in this case the inside and outside regions in fig. A.1(c) will have reversed roles.

3 Derivation of equation 1.20, the Laplace transform of a sampled wave

Assume that $F(s)$ is a rational polynomial in s with partial-fraction expansion given by

$$F(s) = \frac{A_1}{s + s_1} + \frac{A_2}{s + s_2} + \ldots = \sum_i \frac{A_i}{s + s_i} \qquad \text{(A.2)}$$

The inverse Laplace transform gives us the corresponding time function as a sum of exponentials, i.e.

$$f(t) = \sum_i A_i e^{-s_i t}$$

To determine the sampled transform $F_s(s)$, we consider the first term of equation A.2 in the following way:

$$F_1(s) = \frac{A_1}{s + s_1} \to f_1(t) = A_1 e^{-s_1 t}$$

Using the sampled function $f_1 kT) = A_1 e^{-s_1 kT}$ in equation 1.17, we obtain the sampled transform

$$F_{s1}(s) = A_1 \sum_{k=0}^{\infty} e^{-s_1 kT} e^{-skT}$$

or summing this geometric series, we obtain the following closed form

$$F_{s1}(s) = \frac{A_1}{1 - e^{-(s+s_1)T}} \tag{A.3}$$

The poles of this function are given by $e^{-(s_p + s_1)T} = 1$, i.e. $-(s_{pn} + s_1)T = jn2\pi$, or $s_{pn} = -s_1 - jn(2\pi/T)$ with $n = 0, \pm 1, \pm 2, \ldots$. This shows that a pole at s_1, for a continuous time signal, will be repeated along a vertical line, for the sampled signal, at intervals $2\pi/T = \omega_s$, where ω_s is the sampling frequency in rad s^{-1}. We expand now $F_{s1}(s)$ into a sum of partial-fractions

$$F_{s1} = \sum_n \frac{c_n}{s - s_{pn}}$$

where c_n are evaluated from equation A.3 using

$$c_n = \frac{A_1}{\dfrac{d}{ds}[1 - e^{-(s+s_1)T}]}$$

at $s = s_{pj}$, resulting in $c_n = A_1/T$ at each pole.

Therefore, $F_{s1}(s)$ can be written in terms of its poles as

$$F_{s1}(s) = \frac{A_1}{T}\left[\ldots \frac{1}{s + (s_1 - jn\omega_s)} + \ldots \frac{1}{s + (s_1 - j\omega_s)} + \frac{1}{[s + s_1]} \right.$$
$$\left. + \frac{1}{s + (s_1 + j\omega_s)} + \ldots \frac{1}{s + (s_1 + jn\omega_s)} \ldots \right]$$

The same can be done for the other terms of equation A.2 with the result

$$F_s(s) = \frac{A_1}{T}\left[\ldots \frac{1}{s + (s_1 - j\omega_s)} + \frac{1}{s + s_1} + \frac{1}{s + (s_1 + j\omega_s)} + \ldots \right]$$
$$+ \frac{A_2}{T}\left[\ldots \frac{1}{s + (s_2 - j\omega_s)} + \frac{1}{s + s_2} + \frac{1}{s + (s_1 + j\omega_s)} + \ldots \right]$$
$$+ \frac{A_i}{T}\left[\ldots \frac{1}{s + (s_i - j\omega_s)} + \frac{1}{s + s_i} + \frac{1}{s + (s_i + j\omega_s)} + \ldots \right]$$
$$= \frac{1}{T}\left[\ldots F(s - j\omega_s) + F(s) + F(s + j\omega_s) + \ldots \right]$$

or
$$F_s(s) = \frac{1}{T} \sum_{n=-\infty}^{\infty} F(s - jn\omega_s)$$

which is given by equation 1.20. A more rigorous derivation of equations 1.16 and 1.20 can be found, for example, in Freeman(6), chapter 4.

4   Computer programs for nonrecursive (FIR) filters

A FORTRAN program for calculation of Kaiser window coefficients given by

$$w_k(n) = \frac{I_0[2\pi\sqrt{(1-(n/10)^2)}]}{I_0(2\pi)}$$

using $I_0(x) = 1 + \sum_{k=1}^{K}\left[\frac{(x/2)^k}{k!}\right]^2$

```
0              PROGRAM   KCOEFF
1              END
2                        PI = 3.14159265
3                        T = 1. E-09
4                        DO 25 I = 1, 11
5                        G = I - 1
6                        Y = PI*SQRT(1. -0.01*G*G)
7                        E = 1.
8                        DE = 1.
9                        DO 10 J = 1, 100
10                       DE = DE*Y/FLOAT(J)
11                       SDE = DE*DE
12                       E = E + SDE
13                       IF (T - SDE) 10, 10, 20
14             10        CONTINUE
15                       J = 0.
16             20        IF (I .EQ. 1) E0 = E
17                       W = E/E0
18             25        WRITE (2,30) Y, E, W, J
19             30        FORMAT (10X, 3F15.10, I10)
20                       STOP
21                       END
```

A FORTRAN program for calculation of the nonrecursive filter response described by

$$20\log_{10}\frac{|H(m)|}{|H(0)|}$$

where $|H(m)|$ are given by expressions 3.10, 3.15 and 3.18, with results shown in figs 3.4 and 3.6.

```
0              PROGRAM WINDOWTEST
1              END
2                        DIMENSION A(25), SUM(5)
3                        READ (1,10) N
4              10        FORMAT (5X, I2)
5                        DO 70 J = 1, N
6                        READ (1,20) HINC, M, NA
7              20        FORMAT (5X, F5.2, 5X, I5, 5X, I1)
8                        WRITE (2,20) HINC, M, NA
9                        IF (NA .NE. 0) READ (1,25) (A(L), L = 1, M)
10                       WRITE (2,25) (A(L), L = 1, M)
11             25        FORMAT (1X, 5F15.10)
12                       PI = 3.14159265
13             30        H = 0.
14             35        DO 50 K = 1, 5
15                       IF (H .GT. 1) GO TO 70
16                       SUM(K) = 0.25
```

```
17                       DO 40 I = 1, M
18                       G = I
19          40           SUM(K) = SUM(K) + A(I)*COS(G*H*PI)
20                       IF (H .LT. HINC) SUM0 = SUM(1)
21                       SUMN = ABS(SUM0/SUM(K))
22                       SUM(K) = 20.*ALOG10(SUMN)
23                       G = H
24          50           H = H + HINC
25                       WRITE (2,60) G, SUM
26          60           FORMAT (5X, F10.4, 5X, 5F10.3)
27                       GO TO 35
28          70           CONTINUE
29                       STOP
30                       END
```

5   Computer program for recursive (IIR) filters

A FORTRAN program for calculating

$$10 \log_{10} \frac{|H(m)|^2}{|H(r)|^2} \text{ of figs 4.2, 4.4, 4.7, 4.8 and 4.9}$$

where $$|H(m)| = (\text{const}) \frac{A_1 + A_2 \cos m\pi + A_3 \cos 2m\pi + A_4 \cos 3m\pi}{A_5 + A_6 \cos m\pi + A_7 \cos 2m\pi + A_8 \cos 3m\pi}$$

$|H(r)|^2 = |H(0)|^2$ for lowpass filters (figs 4.2, 4.4, 4.7 and 4.8)
$|H(r)|^2 = |H(1)|^2$ for bandpass filter (fig. 4.9), and in this case m represents $2m$.

```
0         PROGRAM DFLTR
1         END
2                       DIMENSION A(8)
3                       PI = 3.14159265
4                       READ (1,10) N
5         10            FORMAT (5X, I2)
6                       DO 30 J = 1, N
7                       READ (1,20) C, A
8         20            FORMAT (1X, F11.8/(4F15.10)
9                       WRITE (2,20) C, A
10                      DO 30 M = 101, 501, 2
11                      AM = M - 1
12                      ANG = AM*PI/100.
13                      TOP = A(1)+A(2)*COS(ANG)+A(3)*COS(2.*ANG)+A(4)*COS(3.*ANG)
14                      DEN = A(5)+A(6)*COS(ANG)+A(7)*COS(2.*ANG)+A(8)*COS(3.*ANG)
15                      HMSQ = C*TOP/DEN
16                      IF (M .EQ. 101) H0SQ = HMSQ
17                      HMSN = HMSQ/H0SQ
18                      HMSN = ABS(HMSN)
19                      IF (HMSN .LE. 0) GO TO 50
20                      RES = 10.*ALOG10(HMSN)
21                      WRITE (2,40) AM, RES
22        30            CONTINUE
23                      STOP
24        40            FORMAT (5X, F5.1, 5X, F9.3)
25        50            WRITE (2,60) AM
26        60            FORMAT (5X, F5.1, 5X, 11HFN INFINITE)
27                      GO TO 30
28                      END
```

The above program is for fig. 4.9, and for figs 4.2, 4.4, 4.7 and 4.8 only step 10 should be changed to: DO 30 M1, 201, 5

6   Mean-square error of the sample mean estimator

Equation 6.5 is obtained from

$$p_e = E\left[\frac{1}{m}\sum_{i=1}^{m} v(i)\right]^2 = \frac{1}{m^2}\sum_{i=1}^{m}\sum_{j=1}^{m} E[v(i)v(j)]$$

$$= \frac{1}{m^2}\sum_{i=1}^{m}\sum_{j=1}^{m} \sigma_v^2 \delta_{ij}$$

where δ_{ij} represents the Kronecker delta, i.e. $\delta_{ij} = 1$ for $i = j$, and $\delta_{ij} = 0$ for $i \neq j$. Therefore, we have

$$\sum_{i=1}^{m}\sum_{j=1}^{m} \delta_{ij} = \sum_{i=1}^{m} (\delta_{i1} + \delta_{i2} + \ldots + \delta_{im})$$
$$= \delta_{11} + \delta_{12} + \ldots + \delta_{1m}$$
$$+ \delta_{21} + \delta_{22} + \ldots + \delta_{2m}$$
$$+ \delta_{m1} + \delta_{m2} + \ldots + \delta_{mm}$$

In the above only δ_{ij} for $i = j$ are equal to 1, all the others are zero. There are m such terms, therefore the double sum in this case reduces to m. The result is then $p_e = \sigma_v^2/m$.

7   Mean-square error for recursive filter

The error is

$$e = \hat{x} - x = -a^m x + (1-a)\sum_{i=1}^{m} a^{m-i} v(i)$$

and the mean-square error is

$$p_e = E(e^2) = E[-a^m x + (1-a)\sum_{i=1}^{m} a^{m-i} v(i)]^2$$

$$= a^{2m} E(x^2) + (1-a)^2 E[\sum_{i=1}^{m} a^{m-i} v(i)]^2$$

since $E[xv(i)] = 0$, because x and $v(i)$ are not correlated.
   Further, we have

$$p_e = a^{2m} S + (1-a)^2 \sum_{i=1}^{m}\sum_{j=1}^{m} a^{m-i} a^{m-j} \sigma_v^2 \delta_{ij}$$

where we have represented $e[v(i)v(j)] = \sigma_v^2 \delta_{ij}$, as in section 6 of the appendix, and we denote $E(x^2)$ as S. Since $\delta_{ij} = 1$ only for $i = j$, the second term simplifies, so we can write

$$p_e = a^{2m} S + (1-a)^2 \sigma_v^2 \sum_{i=1}^{m} (a^{m-i})^2$$

$$= a^{2m} S + (1-a)^2 \sigma_v^2 [a^{2(m-1)} + a^{2(m-2)} + \ldots + a^2 + 1]$$

The sqaure bracketed part is a geometric series, and summing this series we have

$$p_e = a^{2m} S + \frac{(1-a)(1-a^{2m})}{1+a} \sigma_v^2$$

8 Relationship between $a(k)$ and $b(k)$ in section 7.3

We start with $E[e(k)\hat{x}(k-1)] = 0$, and substitute $e(k) = \hat{x}(k) - x(k)$, and using equation 7.11 for $\hat{x}(k)$, we have

$$E\{[a(k)\hat{x}(k-1) + b(k)y(k) - x(k)]\hat{x}(k-1)\} = 0$$

Adding and subtracting $a(k)x(k-1)$, the above becomes

$$E\{[a(k)[\hat{x}(k-1) - x(k-1) + x(k-1)]\hat{x}(k-1)\} = E\{[x(k) - b(k)y(k)]\hat{x}(k-1)\}$$

This can be written as

$$a(k)E[e(k-1)\hat{x}(k-1) + x(k-1)\hat{x}(k-1)] = E\{[1 - cb(k)]x(k) - b(k)v(k)]\hat{x}(k-1)\}$$

where we have used equation 7.9 for $y(k)$.

The first term on the left-hand side, $E[e(k-1)\hat{x}(k-1)] = 0$, because we can write $\hat{x}(k-1) = a(k-1)\hat{x}(k-2) + b(k-1)y(k-1)$ and use the orthogonality equations 7.15 and 7.16 for the previous time $(k-1)$. The second term on the right-hand side $E[v(k)\hat{x}(k-1)] = 0$, because the estimate at time $(k-1)$ is not correlated with the observation noise at time k. Therefore, the above equation reduces to

$$a(k)E[x(k-1)\hat{x}(k-1)] = [1 - cb(k)]E[x(k)\hat{x}(k-1)]$$

Substituting now for $x(k) = ax(k-1) + w(k-1)$ we have $E[w(k-1)\hat{x}(k-1)] = 0$. To show this, we express

$$\hat{x}(k-1) = a(k-1)\hat{x}(k-2) + acb(k-1)x(k-2) + cbx(k-1)w(k-2) + b(k-1)v(k-1)$$

where we have used equations 7.11, 7.9 and 7.6. Averages of all the products of the above with $w(k-1)$ are zero, because all terms are uncorrelated with $w(k-1)$, so we are now left with

$$a(k)E[x(k-1)\hat{x}(k-1)] = a[1 - cb(k)]E[x(k-1)\hat{x}(k-1)]$$

giving
$$a(k) = a[1 - cb(k)]$$

9 Calculation of $b(k)$ and $p(k)$ in section 7.3

Starting with

$$p(k) = E[e^2(k)] = E\{e(k)[\hat{x}(k) - x(k)]\}$$

Substituting for $\hat{x}(k)$ from equation 7.11, and using equations 7.15 and 7.16 we obtain $p(k) = -E[e(k)x(k)]$.

From equations 7.16 and 7.9 we have

$$cE[e(k)x(k)] = -E[e(k)v(k)]$$

which enables the mean-square error to be written as

$$p(k) = \frac{1}{c}E[e(k)v(k)]$$

Substituting for $e(k) = \hat{x}(k) - x(k)$, and using equation 7.11, we obtain an expression for $p(k)$ containing three terms. Two of these, $E[\hat{x}(k-1)v(k)]$ and $E[x(k)v(k)]$, average to zero so we are left with

$$p(k) = \frac{1}{c}b(k)E[y(k)v(k)] = \frac{1}{c}b(k)\sigma_v^2$$

so
$$b(k) = cp(k)/\sigma_v^2$$

To solve the problem completely we return to the mean-square error equation

$$p(k) = E[\hat{x}(k) - x(k)]^2$$
$$= E\{a\hat{x}(k-1) + b(k)[y(k) - ac\hat{x}(k-1)] - x(k)\}^2$$

where we have used equation 7.17. Substituting for $y(k)$ from equation 7.9, and using equation 7.6, we can rewrite the above as

$$p(k) = E\{a[1 - cb(k)]e(k-1) - [1 - cb(k)]w(k-1) + b(k)v(k)\}^2$$

The cross products in the above expression average to zero because $e(k-1)$, $w(k-1)$ and $v(k)$ are independent of each other, so we have

$$p(k) = a^2[1 - cb(k)]^2 p(k-1) + [1 - cb(k)]^2\sigma_w^2 + b^2(k)\sigma_v^2$$

where $p(k-1) = E[e^2(k-1)]$, $\sigma_w^2 = E[w^2(k-1)]$ and $\sigma_v^2 = E[v^2(k)]$. Substituting $p(k) = b(k)\sigma_v^2/c$ from the result derived earlier in this secton of the appendix, we have

$$b(k)\{\sigma_v^2 + c^2[a^2 p(k-1) + \sigma_w^2]\} = c[a^2 p(k-1) + \sigma_w^2]$$

from which the solution, equation 7.18, is obtained. The other solution of the quadratic equation for $b(k)$ if $b(k) = 1/c$. This solution is neglected because it is time-invariant because c is constant, while the first solution is time-varying through $p(k-1)$.

10 FORTRAN program for Kalman filter in section 9.4

```
0          PROGRAM KALMAN
1          END
2                   DIMENSION P(2,2)
3                   DIMENSION Y(6), AK(2)
4                   X = 95.0
5                   V = 1.0
6                   P(1,1) = 10.0
7                   P(1,2) = 0.0
8                   P(2,1) = 0.0
9                   P(2,2) = 1.0
10                  Y(1) = 100.0
11                  Y(2) = 97.9
12                  Y(3) = 94.4
13                  Y(4) = 92.7
14                  Y(5) = 87.3
15                  Y(6) = 82.1
16                  DO 20 I = 1, 6
17                  X = X + V - 0.5
18                  V = V - 1.0
19                  P(1,1) = P(1,1) + P(2,2) + 2.0*P(2,1)
20                  P(2,1) = P(2,1) + P(2,2)
21                  D = P(1,1) + 1.0
22                  AK(1) = P(1,1)/D
23                  AK(2) = P(2,1)/D
24                  Z = Y(I) - X
25                  X = X + AK(1)*Z
26                  V = V + AK(2)*Z
27                  P(2,2) = P(2,2) - AK(2)*P(2,1)
28                  P(2,1) = P(2,1) - AK(2)*P(1,1)
29                  P(1,1) = (1.0 - AK(1))*P(1,1)
30                  WRITE (2,200) Y(I), X, V, P(1,1), P(2,1), P(2,2)
31                  WRITE (2,99)
32          20      CONTINUE
33          200     FORMAT (1X, 10(1H ), 7F15.4)
34          99      FORMAT (1X, 1H )
35                  STOP
36                  END
```

Comments

Steps 17, 18   are based on equation 9.19.
Steps 19, 20   come from equation 8.24 for $Q = 0$, matrix A as in equation 9.19, and denoting

$$P(1) = \begin{bmatrix} P(1,1) & P(1,2) \\ P(2,1) & P(2,2) \end{bmatrix}$$

where $P(1, 2) = P(2, 1)$ because of symmetry.

Steps 21, 22, 23   follow from the above and equation 8.23.

Steps 25, 26   are position and velocity estimates from equation 8.22.

Steps 27, 28, 29   come from equation 8.25 with $P_1(k)$ represented as above in steps 19, 20.

11   State response of a first-order scalar system

A single-input, single-output, first-order, linear, lumped, time-invariant, deterministic (noise-free) dynamical system with forcing input $u(t)$ can be described by a differential equation

$$\dot{x}(t) = fx(t) + gu(t) \tag{A.4}$$

with given $x(0)$. The solution of this equation in the continuous-time is

$$x(t) = e^{ft}x(0) + \int_0^t e^{f(t-\tau)}gu(\tau)\,d\tau \tag{A.5}$$

We are interested in the values of x at discrete points in time, at intervals of T. Equation A.5 allows us to write these values in terms of one before. Setting times successively to $T, 2T, \dots, kT$ we have

$$x(T) = e^{fT}x(0) + \int_0^T e^{f(T-\tau)}gu(\tau)\,d\tau$$

$$= e^{fT}x(0) + \int_0^T e^{f\tau'}gu(T-\tau')\,d\tau'$$

$$x(2T) = e^{fT}x(T) + \int_T^{2T} e^{f(2T-\tau)}gu(\tau)\,d\tau$$

$$= e^{fT}x(T) + \int_0^T e^{f\tau''}gu(2T-\tau'')\,d\tau''$$

General solution is then deduced as

$$x(kT) = e^{fT}x[(k-1)T] + \int_0^T e^{f\tau}gu(kT-\tau)\,d\tau \tag{A.6}$$

where for convenience the transport on the new variable of integration has been dropped. If T is short enough relative to the variations in $u(t)$, so that $u[(k-1)T]$ is a reasonable approximation to $u(t)$ throughout time from $(k-1)T$ to kT, we can write

$$x(kT) = e^{fT}x[(k-1)T] + gu[(k-1)T]\int_0^T e^{f\tau}\,d\tau$$

or $$x(kT) = ax[(k-1)T] + bu[(k-1)T] \tag{A.7}$$

where $$a = e^{fT} = 1 + fT + \frac{(fT)^2}{2!} + \dots \tag{A.8}$$

and $$b = g\int_0^T e^{f\tau}\,d\tau = \frac{g}{f}(e^{fT} - 1) \tag{A.9}$$

With T fixed, a and b are just numbers, easily computed once and for all.

There is an obvious similarity of form between the discrete-time state equation A.7 and the continuous-time state equation A.4. However, A.7 is a *difference* equation and A.4 is a *differential* equation. The particular advantage of equation A.7 is that it is in the form required for digital computation of the state response.

One can also obtain a difference equation similarly as in section 1.1, by using here forward differences in A.4 which produces

$$x(kT) = (1 + fT)x[(k-1)T] + gTu[(k-1)T] \qquad (A.10)$$

This is an approximation to equation A.8 with $e^{fT} \simeq 1 + fT$, used for example in example 9.5.

12 State response of a first-order vector system

The idea of signal and data vectors has been developed in section 8.1. We extend it here by considering a multi-input, first-order, linear, lumped, time-invariant, deterministic (noise-free) dynamical system with forcing input $\mathbf{u}(t)$ which is described by the vector differential equation

$$\dot{\mathbf{x}}(t) = F\mathbf{x}(t) + G\mathbf{u}(t) \qquad (A.11)$$

where $\mathbf{x}$ is the state vector, $\mathbf{u}$ is the forcing vector, F is the state matrix (number of rows = number of columns = number of state variables), G is the excitation matrix (number of rows = number of state variables, and number of columns = number of forcing variables). We could go through a similar procedure, as in appendix 11, by considering first the continuous-time solution, discretizing it, and developing the vector difference equation. However, it is simpler to use results (A.7) to (A.9) for the vector case as follows:

$$\mathbf{x}(kT) = A\mathbf{x}[(k-1)T] + B\mathbf{u}[(k-1)T] \qquad (A.12)$$

where $\quad A = e^{FT} = I + FT + \dfrac{(FT)^2}{2!} + \dots \qquad (A.13)$

and $\quad B\left(\displaystyle\int_0^T e^{F\tau}d\tau \right)G = F^{-1}(e^{FT} - I)G \qquad (A.14)$

Note that A, B, F, G are matrices, and I is the unitary matrix (diagonal terms = 1, others = 0). With T fixed, A and B are computed once and for all. $A = e^{FT}$ is often referred to as the transition matrix denoted usually as $\Phi(T)$. In the case of B in equation A.14 one may need to use the integral if F^{-1} does not exist, i.e. if the determinant of F is zero.

Therefore, it is useful to know that $A = e^{FT}$, and then also B can be obtained by other methods. One is, for example, by means of modal decomposition, Brogan(40), which is based on the eigenvalues and corresponding eigenvectors of the state matrix F. Alternatively, one may obtain $\Phi(t, t_0)$ directly for a given (2×2) state matrix F as given in Liebelt's Appendix(41). The matrix A is then obtained by taking $t_0 = 0$, $t = T$ in $\Phi(t, t_0)$. The solution for B is obtained by the integration indicated in equation A.14.

References

1   Rabiner, L. R. *et al.* Terminology in digital signal processing. *IEEE Trans. Audio and Electroacoustics*, **AU-20**, 322–37, Dec. 1972
2   Steiglitz, K. *An introduction to discrete systems.* John Wiley, 1974
3   Hovanessian, S. A. *et al. Digital computer methods in engineering.* McGraw-Hill, 1969
4   Cadzow, J. A. *Discrete-time systems: an introduction with interdisciplinary applications.* Prentice-Hall, 1973
5   Anderson O. D. *Time series analysis and forecasting: the Box-Jenkins approach.* Butterworth, 1975
6   Freeman, H. *Discrete-time systems.* John Wiley, 1965
7   Jury, E. J. *Theory and applications of the z-transform method.* John Wiley, 1964
8   Robinson, E. A. *Statistical communication and detection.* Griffin, 1967
9   Kaiser, J. F. *Digital filters, systems analysis by digital computer*, Kuo, F. F. and Kaiser, J. F. (editors). John Wiley, 1966
10 Oppenheim, A. V. and Schaffer, R. W. *Digital signal processing.* Prentice-Hall, 1975
11 Proakis, J. G. and Manolakis, D. G. *Digital signal processing*, 2nd edition. Macmillan, 1992
12 Skolnik, M. I. *Introduction to radar systems.* McGraw-Hill, 1962
13 Rabiner, L. R. and Gold, B. *Theory and applications of digital signal processing.* Prentice-Hall, 1975
14 Linke, J. M. Residual attenuation equalization of broadband systems by generalized transversal networks. *Proc. IEE*, **114**, 3, 339–48, 1967
15 Parks, T. W. and Burrus, C. S. *Digital filter design.* John Wiley, 1987
16 Kaiser, J. F. *Digital filters: systems analysis by digital computer*, Kuo, F. F. and Kaiser, J. F. (editors). John Wiley, 1966
17 Kuc, R. *Introduction to digital signal processing.* McGraw-Hill, 1988
18 Bogner, R. E. and Constantinides, A. G. (editors). *Introduction to digital filtering.* John Wiley, 1975
19 Haykin, S. A unified treatment of recursive digital filtering. *IEEE Trans. on Automatic Control,* **17**, 113–116, Feb. 1972
20 Smith, D. A. *et al.* Active bandpass filtering with bucket-brigade delay lines. *IEEE Journal of solid-state Circuits*, **SC-7**, 5, 421–5, Oct. 1972
21 Bozic, S. M. Chapter 7 in *Digital signal processing*, N. B. Jones (editor). IEE Control Engineering Series, Vol. 22, Peregrinus, 1982
22 DeFatta, D. J. *et al. Digital signal processing: a system design approach.* John Wiley, 1988
23 Antoniou, A. *Digital filters*, second edition. McGraw-Hill, 1993.
24 Bozic, S. M. Hardware realization of wave-digital filters. *Microelectronics Journal*, **22**, 1991, 101–108.
25 Tan, E. C. Variable lowpass wave-digital filters. *Electron. Lett.*, 324–326, 15th April, 1982
26 Abramowitz, M. *et al.* (editors). *Handbook of mathematical functions.* Dover Books, 1965
27 Schwartz, M. and Shaw, L. *Signal processing: discrete spectral analysis, detection and estimation.* McGraw-Hill, 1975
28 Van den Enden, A. W. M. *Discrete-time signal processing.* N. A. M. Verhoeckx, Prentice-Hall, 1989
29 Chrochiere, R. E. and Rabiner, L. R. *Multirate digital processing.* Prentice-Hall, 1983
30 Bozic, S. M. Wiener IIR filter design. *Int. J. Electronics*, **58**, 3, 463–470, 1985
31 Candy, J. V. *Signal processing, the moden approach.* McGraw-Hill, 1988, ch. 8
32 Lim, J. S. and Oppenheim, A. V. *Advanced topics in signal processing.* Prentice-Hall, 1988, ch. 5
33 Kalman, R. E. A new approach to linear filtering and prediction problems. *Trans ASME, J. of Basic Engineering*, 35–45, March 1960

34 Sorenson, H. W. Least-squares estimation from Gauss to Kalman. *IEEE Spectrum*, **7**, 63–8, July 1970

35 Sorenson, H. W. Kalman filtering techniques. *Advances in control systems*, 3, C. T. Leondes (editor), Academic Press, 1966

36 Sage, A. P. and Melsa, J. L. *Estimation theory with applications in communications and control*. McGraw-Hill, 1971

37 Jazwinski, A. H. *Stochastic processes and filtering theory*. Academic Press, 1970

38 Candy, J. V. *Signal processing: the model based approach*. McGraw-Hill, 1986

39 Singer, R. and Behnke, K. Real-time tracking filter evaluation and selection for tactical applications. *IEEE Trans. Aerosp. Electron. Syst.*, **AES-7**, 100–110, Jan. 1971

40 Brogan, W. L. *Modern control theory*. Quantum Publishers, 1974

41 Liebelt, P. B. *An introduction to optimal estimation*. Addison-Wesley, 1967

Index